The Neuromodulation Casebook

The Neuromodulation Casebook

Edited by

Jeffrey Arle
Associate Professor of Neurosurgery,
Harvard Medical School,
Boston, MA, United States

Associate Chief of Neurosurgery,
Beth Israel Deaconess Medical Center,
Boston, MA, United States

ACADEMIC PRESS

An imprint of Elsevier

Academic Press is an imprint of Elsevier
125 London Wall, London EC2Y 5AS, United Kingdom
525 B Street, Suite 1650, San Diego, CA 92101, United States
50 Hampshire Street, 5th Floor, Cambridge, MA 02139, United States
The Boulevard, Langford Lane, Kidlington, Oxford OX5 1GB, United Kingdom

Notices
Knowledge and best practice in this field are constantly changing. As new research and experience broaden our understanding, changes in research methods, professional practices, or medical treatment may become necessary.

Practitioners and researchers must always rely on their own experience and knowledge in evaluating and using any information, methods, compounds, or experiments described herein. In using such information or methods they should be mindful of their own safety and the safety of others, including parties for whom they have a professional responsibility.

To the fullest extent of the law, neither the Publisher nor the authors, contributors, or editors, assume any liability for any injury and/or damage to persons or property as a matter of products liability, negligence or otherwise, or from any use or operation of any methods, products, instructions, or ideas contained in the material herein.

Library of Congress Cataloging-in-Publication Data
A catalog record for this book is available from the Library of Congress

British Library Cataloguing-in-Publication Data
A catalogue record for this book is available from the British Library

ISBN: 978-0-12-817002-1

For information on all Academic Press publications visit our website at
https://www.elsevier.com/books-and-journals

Publisher: Nikki Levy
Acquisition Editor: Natalie Farra
Editorial Project Manager: Pat Gonzalez
Production Project Manager: Surya Narayanan Jayachandran
Cover Designer: Mark Rogers

Working together
to grow libraries in
developing countries

www.elsevier.com • www.bookaid.org

Typeset by TNQ Technologies

Contents

Section IV
Vagus nerve stimulation and responsive neurostimulation

Contributors

Jeffrey E. Arle, Neurosurgery, Harvard Medical School, Boston, MA, United States; Neurosurgery, Beth Israel Deaconess Medical Center, Boston, MA, United States

Jianwen Wendy Gu, Principal Field Engineer, Boston Scientific Neuromodulation, Valencia, CA, United States

Laura Lombardi-Karl, Anesthesiology, Critical Care, and Pain Medicine, Beth Israel Deaconess Medical Center, Boston, MA, United States

Alon Mogilner, Neurosurgery and Anesthesiology, NYU Langone Medical Center, New York, NY, United States

Musa Moris Aner, Anesthesiology, Critical Care, and Pain Medicine, Beth Israel Deaconess Medical Center, Boston, MA, United States

Mohamed Osman, Anesthesiology, Critical Care, and Pain Medicine, Beth Israel Deaconess Medical Center, Boston, MA, United States

Timothy Ray Deer, The Center for Pain Relief, Inc., Charleston, WV, United States

Patricia Osborne Shafer, Health Information and Resources, Programs and Services, Epilepsy Foundation, Landover, MD, United States; Neurology, Beth Israel Deaconess Medical Center, Boston, MA, United States

Jay L. Shils, Anesthesiology, Intraoperative Neurophysiologic Monitoring, Rush University Medical Center, Chicago, IL, United States

Tom Simopoulos, Department of Anesthesiology, Critical Care Medicine and Pain Medicine, Division of Pain Medicine, Beth Israel Deaconess Medical Center, Boston, MA, United States

Cyrus Yazdi, Department of Anesthesiology, Critical Care Medicine and Pain Medicine, Division of Pain Medicine, Beth Israel Deaconess Medical Center, Boston, MA, United States

Preface

Neuromodulation is a very current, cutting-edge field and encompasses a variety of techniques and therapies, ranging from deep brain stimulation for a multitude of disorders to spinal cord stimulation, peripheral nerve stimulation, cortical stimulation, and cranial nerve stimulation, not to mention a collection of growing noninvasive therapies and other implanted types of devices that interface with the nervous system. It overlaps the boundaries of functional neurosurgery, spine surgery, pain management, and bioengineering. I have been happily involved in the field of neuromodulation for over 20 years and can see there is a need for a type of "mini-fellowship" where practitioners who want to start or add to their repertoire of neuromodulatory therapies can find all the basics from expert mentors in one convenient location. This book seeks to fill that need by putting in a single book all the practical, hands-on decision-making using realistic case examples—how they are managed and how variations in this management would be handled.

All neuromodulation therapies involve a host of decisions regarding patient selection, target, technical aspects, the technique itself, programming, and follow-up care, as well as complication management. Complication management alone, in fact, is critical in most surgical or procedural therapies and drives the structure and depth of most training programs. As many surgeons might recall hearing from their own mentors, anyone can put a device in, but what is important is how to decide in whom to place the device and then in managing the situation when things do not go as planned in placing or following the patient and the device.

This book is case-based, for the simple reason that people learn best with case-based examples in these types of fields. As a book presents with a format and size, it cannot possibly do justice to every nuance and type of therapy without becoming uninteresting and unwieldy; however, the format follows a single case but pauses periodically to review typical variations that could have arisen at that juncture of the care. Answers are succinctly given to each of these variations, and then the case timeline resumes. Overall, the goal with this format is to have the reader arrive at the end of each chapter, and at the end of the book itself, with a feeling that they have seen not only the typical cases and presentations and problems but also a large swath of the variations that can occur in real practice, as well as how to manage them. In a sense, this is what a fellowship training year would accomplish—but it is rare to find a fellowship that would allow one to gain expertise in all of these areas of neuromodulation since they span so many individual subfields (e.g., pain management, movement disorders, epilepsy). Most likely, a stereotactic and functional fellowship will only cover one or two of these areas well.

As noted, the cases cover these most typical areas where neuromodulation is used—deep brain stimulation for tremor, Parkinson disease, and dystonias; spinal cord stimulation for failed back surgery and complex regional pain syndromes; peripheral nerve stimulation for pain in a variety of targets; vagus nerve stimulation for epilepsy. This retains a practical grounding for the book—like a handbook and a guidebook, and a personal mentoring all in one, without straying too far into areas that only come up rarely if ever for the majority of clinicians in these fields.

Finally, I have tried to include examples that involve not just surgeons per se but also nonsurgeon proceduralists: the physiatrist who does spinal cord stimulator trials, for example, or the pain management physician who places peripheral nerve leads using ultrasound guidance or face leads. It also includes expert insight into programming these devices, in a similar type of format. In its ultimate structure, it is more like reading a series of good short stories than a pedantic, bullet point—ridden listing of references or "tips" per se, though no doubt there is still some of that. My goal is to allow the reader to find real examples of decisions they need to make every day and a frank discussion from those who are considered "leaders" in the field on how they work through them. In addition, the text serves as a companion (to *Essential Neuromodulation* and *Innovative Neuromodulation*) for a course in Biomedical Engineering or a reasonable reference for the early training of fellows in pain programs and functional neurosurgeons.

Section I

Spinal cord stimulation

Chapter 1

Percutaneous trial lead placement for failed back surgery syndrome

Laura Lombardi-Karl, Musa Moris Aner

Anesthesiology, Critical Care, and Pain Medicine, Beth Israel Deaconess Medical Center, Boston, MA, United States

Scenario

A 57-year-old female with a history of L4-L5 and L5-S1 laminectomy and fusion presents with persistent axial back and left leg pain. She has been diagnosed with failed back surgery syndrome (FBSS). She has trialed several medial branch blocks and lumbar epidural steroid injections with minimal relief. She is currently taking gabapentin 300 mg three times daily and has participated in several courses of physical therapy with minimal benefit. She is anxious that no solutions will be found for her chronic pain and seems willing to try alternative therapies. Discussion takes place to consider spinal cord stimulation and the possibility of performing a trial of the stimulation to see if it can help her. She expresses interest in proceeding and is given some literature to review on the therapy.

> **Variation**: *Conservative multimodal management has not been trialed* – Spinal cord stimulation is generally considered after patients have failed more conservative management with medication, physical therapy, or other less invasive interventional techniques, although this is not necessarily required. Placement of SCS should not be viewed as a replacement or contraindication to multimodal analgesia, especially considering that patients often have multiple pain generators and pain is often multifactorial. In many cases, patients will continue taking adjuvant medications and pursuing physical therapy. However, there should be an effort to reduce opioid consumption, and a plan for opioid reduction should be discussed with the patient as part of preoperative planning.

The Neuromodulation Casebook. https://doi.org/10.1016/B978-0-12-817002-1.00001-8

3

Variation: *The patient has had multiple surgeries at different levels –* It is important to confirm the level of a patient's prior surgeries as this may interfere with safe placement of percutaneous leads. Laminectomies will result in a disrupted ligamentum flavum, while fusion hardware may interfere with imaging. Prior surgery can result in epidural scarring that extends beyond the level of surgery. This can interfere with the loss of resistance technique and pose difficulty threading leads. Severe central canal stenosis may be a contraindication for an SCS as placement of a space-occupying lead may cause myelopathy. Although this is of greater concern in cervical SCS placement, it should be considered at all levels. Before SCS implant, patients with FBSS should be reevaluated by their surgeon to ensure there are no structural issues that would require further surgery. It is also valuable to obtain updated imaging with contrast to assess for new or worsening pathology. Thoracic spine magnetic resonance imaging (MRI) can also be included as part of preoperative planning for thoracic SCS placement.

Variation: *The patient is elderly and has mild cognitive impairment but good family support –* Although family members help the patient to remember and effectively communicate at visits, using the device and communicating eventual needs for reprogramming and functional improvements may be hampered. At times, the patient's underlying pain state or comorbidities may influence the surgeon's choice of SCS manufacturer. Device selection is a complex process taking into account the characteristics of the patient's pain, surgeon's preference, and patient's ability to communicate. Patients with anxiety disorders or difficulty describing their pain may benefit from placement of a device that does not require extensive intraoperative mapping and patient cooperation, such as high-frequency 10 kHz. All elderly patients should go through psychiatric evaluation as standard of care, not necessarily to rule out placement but to find ways to optimize the patient for a better outcome.

Variation: *The patient has a history of breast cancer that has been in remission for 2 years –* Patients with medical conditions requiring frequent MRI imaging, such as cancer, may influence the choice of SCS device. Some SCS devices are fully MRI compatible, some are conditionally compatible, while some are not compatible at all. A full discussion with the patient's oncologist or other relevant providers regarding the likely need for future MRI imaging should occur before SCS placement. Similarly, there may be less utility in selecting an MRI-compatible device for patients who already have a contraindication to MRI, such as those with non−MRI-compatible pacemakers or deep-brain stimulators. Other patients who may have an increased need for MRI imaging include those with a concern for intracranial pathology, such as patients with a history of seizures or headache, as well as patients with multiple joint complaints or surgeries.

Scenario (continued)

A full physical examination is completed, and no new neurological deficits are noted. The patient's MRIs and relevant laboratory test results are reviewed, and she has had an unremarkable preoperative psychological assessment. The patient expresses reasonable expectations and understands SCS trial restrictions. The decision is made to proceed with a percutaneous SCS trial.

Variation: *The patient reports that she expects her pain to resolve after SCS placement* – It is important to consider both disease- and patient-specific factors when considering SCS. FBSS is an indication with high probability of success with SCS. Newer technology has enabled pain states traditionally resistant to SCS, such as axial low back pain, to have better response. Proper patient education and expectation setting is key for success and reduction of placebo, as well as false positives. Before trial, the potential range of outcomes should be discussed with the patient. Patients with unrealistic expectations, such as anticipation of complete pain relief, those who seem hesitant to proceed, and any patient with uncontrolled anxiety, depression, addiction, or other psychiatric issue, should be viewed with caution. We recommend all patients meet with a psychiatric health-care provider before SCS trial for a thorough evaluation of potential psychiatric barriers. Patients with a history of noncompliance with other treatment, unstable social or health insurance situations, and those with limited cognitive ability require thoughtful evaluation before proceeding. In addition, clinicians and patients should discuss possible differences in relief provided by the trial versus permanent implant. Ideally, this conversation should have been had at multiple times by multiple providers, including the surgeon and pain psychologist.

Variation: *The patient is on anticoagulation for atrial fibrillation* – Anticoagulation continues to be a barrier in SCS placement. Multiple guidelines exist; American Society of Regional Anesthesia (ASRA) guidelines are the most frequently used ones. Following these guidelines assumes that placement of an epidural infusion catheter is similar to placement of an SCS lead. Risk of stopping anticoagulation for SCS placement requires risk stratification and a thorough discussion with the patient's neurologist, cardiologist, or hematologist. Although physicians in other specialties may be familiar with surgical techniques, they may not be aware of the anticoagulation holding requirements for an SCS trial. Not only does anticoagulation need to be held in advance of the trial, it needs to be held for the duration of the trial. One important consideration is that although SCS trial and permanent placement may require holding anticoagulation for a relatively longer duration, this is only for two discrete intervals. After SCS placement, patients should then require fewer or no other injections, reducing the overall number of times anticoagulation will need to be held. The appropriate time to restart anticoagulation should also be discussed with the patient's multidisciplinary team. There is no consensus regarding the use of paddle versus percutaneous leads for permanent placement. All patients should be instructed to hold nonsteroidal anti-inflammatory drugs for 7 days, and then it is held for the duration of the trial. Exceptions to this include the selective Cox2 inhibitors.

Variation: *The patient is on chronic antibiotics for urinary tract infection prophylaxis, with no active signs of infection* – Active infection is a contraindication to proceeding with trial or permanent lead placement. We recommend patients be off of all antibiotics for 10 days before the procedure. Patients chronically on antibiotics (i.e., UTI prophylaxis) may continue. Perioperative antibiotics (cefazolin or clindamycin in patients allergic to penicillin) are recommended. We do not recommend routine use of postoperative antibiotics. Intraoperatively, strict adherence to a sterile technique is paramount, and a proper surgical technique is the first-line defense against infection. Vancomycin powder applied intraoperatively or IPG anti-microbial pouches could be considered. We recommend all patients have preoperative Methicillin-resistant Staphylococcus aureus (MRSA) screening before permanent implant; however, we do not advocate regular MRSA screening before an SCS trial. Patients who test positive are treated with intranasal mupirocin for 5 days preoperatively.

Scenario (continued)

Several weeks later, the woman returns to the operating room for the scheduled procedure. All preoperative assessments have been completed, and the procedure begins with the patient receiving mild sedation and placed in the prone position with cushioning and pillows so that she feels comfortable. She is awake and alert enough to readily answer questions and confirm issues throughout the procedure, and this status is checked frequently. The region for percutaneous needle entry is assessed with fluoroscopic imaging and infiltrated with 0.5% bupivacaine for comfort. The needle is placed and slowly advanced with a loss of resistance technique. The lead is placed though the needle after removal of the needle stylet. In short order, the lead is advanced along the midline toward the T7-8 level with small deviations corrected along the way. Only a short interlude of backing the electrode down and readvancing is necessary.

Stimulation begins once the lead is in a reasonable position and she reports that she has paresthesia feelings in the low back region and the left leg more than the right but does feel some coverage in the right leg lower down and in the foot slightly. This seems reasonable to work with for trial information, and a decision is made to leave only this one lead in place. Impedances are within reasonable ranges. The lead is then brought out through the opening, and a strain-relief loop is secured on the skin with Tegaderm. The end of the lead is left free and accessible for the later attachment to the portable stimulation device provided by the company for programming and stimulation. She is allowed to recover from the sedation, and after being able to walk to the bathroom on her own with an assistant near and void on her own, she is discharged home with a small amount of medication for pain management.

> **Variation**: *The patient is diabetic, and on the day of surgery, the patient presents with a glucose level of 250 mg/dL* – There is no standard consensus on perioperative glucose management of diabetic patients, although individual institutions may have their own guidelines. Patients with HbA1c >9% may benefit from having their procedure delayed until better glucose control can be achieved. In the immediate perioperative period, blood glucose level >200 mg/dL may be managed with IV insulin. SCS is an elective procedure, and patients should be optimized before proceeding with surgery. Hyperglycemia in the perioperative period is associated with an increased risk of surgical site infection, delayed wound healing, and sepsis. However, an additional consideration is that patients with poorly controlled glucose may not be candidates for future steroid injections, limiting their analgesic options. Moreover, patients may be able to be more physically active and achieve better glucose control once pain is better controlled.

> **Variation**: *The lead is difficult to get into midline position and deviates too far right or left* – This is one of the most common and frustrating intraoperative issues. There are numerous reasons for difficult lead advancement. Accessing the epidural space too far lateral may make it difficult to advance on a midline trajectory. Patient anatomy may present challenges as well. Patients with scoliosis, scar tissue from prior surgeries, and plica mediana dorsalis may present challenges to lead placement. Fortunately, these issues can often be overcome with troubleshooting techniques. The use of a coude-tipped needle instead of

a straight epidural needle can help to access and maneuver the lead in the epidural space. Some manufacturers provide different stylets which can be trialed. Placing a slight bend, up to 30°, on the tip of the stylet lead may provide greater maneuverability. Withdrawing the lead into the needle past the bevel and adjusting the orientation of the needle tip may help to set the lead on a better trajectory when it is advanced. We advocate the use of advancing the lead under live fluoroscopy. Low-dose fluoroscopy can be used to reduce radiation exposure. If the patient complains of paresthesia during advancement, the surgeon should stop advancing the lead and check a lateral image to see if the lead has advanced into the lateral gutter or ventral epidural space. In our practice, trials are carried out under conscious sedation or local anesthesia, where surgeon-patient contact is continuously maintained and surgeons can respond in real time to patient complaints. The use of a guidewire to establish a straight midline path may be employed; however, caution must be exercised as dural tears or damage to the surrounding structures can result. We do not advocate using force to push past an obstruction. If the lead still will not advance midline despite adjustment, withdrawing the needle and redirecting using the loss of resistance technique may be necessary. If the lead still cannot be advanced, entering at a different level may be required. Ideally, the lead will advance in the posterior midline position and follow a straight path. However, sometimes, steering can follow a curved or s-shaped path depending on plica or scarring and still result with the lead in an appropriate position.

Variation: *An inadvertent dural puncture occurs at the preferred entry space* – Inadvertent dural puncture can occur during percutaneous lead placement, either via epidural needle or via lead placement. Patient risk factors include prior surgery at the site of needle entry, obesity, spinal stenosis, and patient movement. Frequent imaging, use of the contralateral oblique view (CLO) view, and frequent flushing to prevent a blood clot from obstructing the lumen of the Tuohy needle may help to prevent a dural puncture. If a dural puncture occurs, it is at the discretion of the surgeon to decide whether to proceed with the procedure or abort. Inadvertent dural punctures present several potential complications. Puncture with 14g Tuohy or a larger size needle can cause significant damage to the dura and result in a range of complications including postdural puncture headache (PDPH). PDPH symptoms include positional headache, diplopia, tinnitus, photophobia, and neck pain. PDPH may be managed conservatively with rest, hydration, and caffeine. An epidural blood patch may be considered if symptoms do not resolve. Epidural blood patch should be used with caution because of the risk of placing a culture medium around a foreign body; however, blood patches have been performed successfully in patients with cerebrospinal fluid (CSF) leaks after SCS implantation. If possible, blood patch should be performed at the level below the dural puncture. Prophylactic blood patches are not recommended at the time of the trial. One important consideration is that patients with PDPHs will be unable to perform their activities of daily living and will be unable to assess the efficacy of an SCS trial. If the decision is made to proceed, the needle is removed and the epidural space is reapproached at a different level, likely one level above, on the ipsilateral side or on the contralateral side. If one lead has been placed successfully before the dural puncture, aborting the procedure in favor of a single-lead system may be elected depending on coverage. If the decision is made to proceed after a dural tear, the operator can opt to keep the patient on antibiotic for the duration of the trial. Additionally, CSF in the epidural space can change conductance and interfere with stimulation.

Variation: *Using a traditional paresthesia technique, we were unable to map patient's pain* – Prior studies have identified patterns of distribution for different areas of pain. If traditional placement does not provide adequate coverage of the targeted area, the leads may need to be placed at higher segments and tested patiently segment-by-segment until proper stimulation can be obtained. Anatomical midline may not correspond exactly to physiological midline, and lead placement may need to be adjusted depending on the pattern of stimulation. In high-frequency techniques, placement is mostly anatomical and does not need paresthesia mapping for trial information.

Variation: *The patient complains of paresthesia in the trunk and stomach* – If the area of paresthesia does not correspond to the area of the patient's pain, varying the pulse width or amplitude should be attempted. If a successful program cannot be found, it may be necessary to reposition one or both leads. Paresthesias in the trunk suggest that the lead is too far lateral or ventral and needs to be repositioned.

Variation: *The patient is too sedated to participate in intraoperative paresthesia mapping* – Ensuring that the patient is sufficiently alert, cooperative, and able to communicate with the surgical team is essential to successful intraoperative stimulation testing. Before the procedure, the surgeon should discuss the need for periods of intra-operative awareness with the anesthesia team, depending on whether traditional or high-frequency stimulation is used. Long-acting sedatives and analgesics, such as ketamine or dilaudid, should be avoided. Patients with limited ability to communicate or those who may be extremely sensitive to anesthetics, such as elderly patients, may benefit from the use of a device that does not require extensive intraoperative stimulation testing for proper placement. Technically, a trial can be completed using only local anesthetic.

Variation: *Appropriate impedances cannot be obtained after lead placement* – Once leads are in place, impedance should be checked, and proper anteroposterior/lateral visualization should be obtained to confirm posterior epidural placement. If the impedances do not match proper placement, anatomical placement of the leads should be reevaluated. Low impedances might infer fluid collection in the epidural space or intrathecal placement of the leads.

Variation: *After confirmation of placement, the lead is inadvertently drawn caudad as the epidural needle is removed* – After confirmation of lead placement, the epidural stylet should be removed under continuous fluoroscopic guidance to maintain proper lead position. This is a time for vigilance to prevent inadvertent lead migration. If the leads move inadvertently, they can be restyletted and advanced to a proper position, unless they are already outside the epidural space.

Variation: *The patient has a Tegaderm allergy* – Securing the leads properly at the end of the procedure is essential to the success of a trial and to preventing infection. To secure the leads to the skin, we recommend applying Dermabond to the lead and skin, then placing Steri-Strips opposite each other with the adhesive toward the lead. We recommend covering this with sterile gauze and Tegaderm. In patients with Tegaderm allergy, an OPSITE dressing may be used. Alternatively, a StayFIX dressing can be applied on the skin with the lead sutured on top. In cases where there is significant concern for potential lead migration, it is possible to use an 18 gauge angiocatheter to tunnel the lead a few centimeters under the skin so that it exits at a less acute angle and is less prone to movement. Caution should also be exercised when flipping the patient from a prone to supine position at the end of the procedure.

Scenario (continued)

The woman is asked to call the company representative at any time (or the office if during normal hours) if there is a concern about the device or a need for reprogramming. This is very important, and it cannot be overemphasized that the patient needs to feel that they should contact the representative or provider right away during the trial period so as to maximize the information obtainable during the trial. She agrees and is brought home where she engages in typical light activities without significant bending or reaching and twisting.

She returns on day 6, having had only one brief visit in the office with the representative to tweak some settings, to get more back coverage to the left side. The dressing is removed and the lead gently pulled out through the skin and a band-aid applied at the opening. There is no evidence of any infection.

She describes being very impressed with the amount of pain benefit where she felt the paresthetic sensations. Overall, she felt that she could sleep better when she turned it down a little as well. Her reported pain relief was about 75%, and she is looking forward to scheduling permanent placement of the device in the near future.

> **Variation**: *The patient experiences significant postoperative pain and cannot tell whether or not the trial is effective* – Patients should be counseled preoperatively that some postprocedural discomfort is to be expected. Patients should continue on their current analgesic regimen during the trial, with the exception of NSAIDS. Postprocedural pain should be significantly improved two to 3 days after the procedure. Oxycodone or hydrocodone PRN may be prescribed for the first few days after the procedure. Using additional analgesics for the duration of the trial should be avoided to prevent a false-positive trial.

> **Variation**: *The patient experiences excellent relief for 2 days after trial lead placement but slips while getting in her car and now reports loss of coverage and recurrence of pain* – An X-ray should be obtained if it appears that a physical dislodgement of the lead has occurred, allowing everyone to know whether the lead has or has not actually displaced. Loss of coverage is a common complication during percutaneous trials. Apart from apparent disconnects on the external parts of the leads, the likely culprit is caudal lead migration. Rarely the lead will migrate cephalad. If there is concern for lead migration during the trial, the patient should be examined and X-rays obtained to confirm proper lead placement. If migration is detected, it may be possible to reprogram the device to recapture appropriate coverage. If the lead has migrated too far, it may be necessary to remove the lead and repeat the trial.

> **Variation**: *The patient reports fevers and increased pain at the trial lead placement site and weakness in the lower extremities* – Although rare, trial lead infections have been reported and need to be treated immediately. Patients with possible infection require urgent evaluation. Signs and symptoms of infection include fever, pain at the site of lead placement, tenderness to palpation, erythema, induration, purulence, and fever. If infection is suspected, leads should be pulled and radiological imaging obtained to evaluate for a possible collection and superficial or deep infection. Patients should be started on antibiotics. A

small group of practitioners advocate the use of postoperative antibiotics during the trial period for prophylaxis. The usual choice of antibiotics is cephalexin twice a day (BID) or three times a day (TID).

<u>Variation</u>: *The patient returns to clinic and reports only 35% relief* – Trial results with <50% improvement are suboptimal and should be interpreted as failure of the device to have a significant impact on the patient's pain. If there is opportunity for reprogramming, the trial can be extended. The standard of care for an SCS trial is 5−7 days, but this can be extended to 10−12 days if there are no signs or symptoms of infection. Sometimes changing stimulation modalities can salvage a trial. If a patient does not have favorable results with the initial stimulation program, for example, high-frequency 10 kHz, this can be transitioned to a different program, such as burst, with possible better results. Different manufacturers have adaptive technology that can be used to transition from one type of stimulation to another.

Chapter 2

Percutaneous trial case for complex regional pain syndrome

Tom Simopoulos, Cyrus Yazdi

Department of Anesthesiology, Critical Care Medicine and Pain Medicine, Division of Pain Medicine, Beth Israel Deaconess Medical Center, Boston, MA, United States

Scenario

A 45-year-old female with a 2-year history of right lower extremity complex regional pain syndrome I (CRPS I) is referred by a community neurologist for a trial of spinal cord stimulation (SCS). The patient incurred an Achilles tendon injury at work that necessitated surgical repair. Several weeks after the surgery, the patient continued to have excessive burning pain with associated swelling, persistent erythema, and piercing cold dysesthesias that encompassed the entire lower extremity involving the foot and calf up to the knee. She did extensive casting followed by physical therapy. Despite adequate healing, there is no real improvement in the intensity and nature of her pain. The extremity was well perfused, and the lack of an explanation for persistent pain led to a diagnostic triple phase bone scan that supported the suspected diagnosis of CRPS. She then received a series of lumbar sympathetic blocks as well as medications in conjunction with physical therapy. The patient derived very short-term relief and complained of poor sleep. In addition, allodynia of the right lower limb meant that she could not even have a bed sheet contact it.

> **Variation**: *The patient has no diagnostic tests before referral* – The diagnosis of CRPS continues to be based on clinical criteria (so-called Budapest Criteria). For medical and as well as future legal matters that not uncommonly follow these patients, it is important to have documented diagnostic criteria that support the CRPS diagnosis. Diagnostic tests such as inflammatory markers, X-rays, magnetic resonance imaging studies, and bone scans can help exclude other causes and support the diagnosis. Once there is a region (nondermatomal) of the extremity that is affected on clinical assessment, then CRPS is evaluated using the

The Neuromodulation Casebook. https://doi.org/10.1016/B978-0-12-817002-1.00002-X

following criteria: sensory alterations (allodynia, hyperalgesia), vasomotor (temperature difference of 1°C, color variation), sudomotor/edematous changes, and motor/trophic alterations (ankylosing of joints, atrophy, dystonia). These criteria are applied in terms of symptomatic complaints in which case there must be three of the categories fulfilled as well as two distinct categories on physical examination. Not uncommonly, with progression of time, the physical findings can diminish in many patients, but others can remain with chronic edema, lower limb atrophy, and loss of range of motion in the toes and ankle. Patients with CRPS, who present for SCS therapy, usually do so after 1–2 years of poorly managed symptoms.

Variation: *There was no response to lumbar sympathetic blockade* – The response to sympathetic blockade that eliminates a substantial component of the patient's pain implies sympathetically mediated pain (SMP). Sympathetic blocks are well established for treatment of the acute phase of CRPS and can help facilitate rehabilitation. Once CRPS establishes a more persistent state, the analgesic effects of the block are not uncommonly short-lived. SCS has become well established as a therapy that can modulate the sympathetic nervous system in both preclinical as well as clinical studies. SCS is therefore highly likely to alleviate the pain in a patient with SMP, but because of the multiple mechanisms of action involved in SCS, the therapy can also have a significant chance of success in sympathetically independent pain.

Variation: *The patient had a trial of gabapentin only* – The management of symptoms using medications for CRPS continues to evolve. The use of membrane-stabilizing agents such as gabapentin, pregabalin, or topiramate may offer relief of neuropathic symptoms. Not uncommonly, the relief is modest, and antidepressant medications in the categories of serotonin and norepinephrine reuptake inhibitors or tricyclics are used to enhance analgesia. Antiinflammatory agents may help with the ankylosing component of the syndrome. Corticosteroids are of use only early in the syndrome. Atypical opioids such as tramadol or tapentadol are not well studied in CRPS but can offer analgesia. The current trend is to avoid traditional mu agonist opioids or at least maintain low dose. Finally, the use of ketamine infusions has been established in randomized control trials to offer short-term relief, but insurance coverage is not uncommonly an issue.

Variation: *In addition, the patient has complaints of right lower extremity weakness, stiffness, and spasm* – Patients with CRPS frequently have additional symptomatic complaints in the extremity that go beyond neuropathic pain and are not likely to be palliated by

SCS. Therefore, a directed discussion with the patient on what SCS will likely treat is important before the trial as well as full implantation. The reversal of dystonia, tremors, muscle spasm, and joint ankylosis are not reliably responsive to SCS therapy. Alleviation of the neuropathic component can still play a major role to facilitate rehabilitation that may permit more symptomatic control of the trophic changes. Additionally, patients with CRPS may have multiple systemic complaints that are not ameliorated or caused by SCS therapy but do nonetheless impact the quality of life for these individuals. Multiple organ systems may be affected that can cause neurocardiogenic syncope, incontinence, feeling of shortness of breath, dysphagia, irritable bowel syndrome, and reflux disease.

Variation: *The patient has high levels of anxiety* – CRPS is a very distressing syndrome that carries a very high disability rate in the chronic state. Patients lose a great deal of autonomy and self-worth that can exacerbate or lead to worsening anxiety and depressive states. In a multidisciplinary pain center, these patients are referred to a psychologist who specializes in the treatment of these conditions as they relate to CRPS. The key issue is to manage the psychopathology to a level that will permit a trial of SCS therapy. A trial of SCS in the presence of high anxiety levels is often indeterminate and confounded by procedural pain and preoccupation with multiple somatic complaints with the inability to focus on the temporary trial. There is a steep learning curve of the technology, and patients need to be able to familiarize themselves with the therapy to maximize their chances for relief. The patient may, out of pure desperation, demand the therapy even if the trial was of unclear benefit.

Scenario (continued)

This woman is counseled that initially she must have a psychological assessment before attempting a trial of stimulation. She agrees, and after this assessment by a qualified professional well versed in evaluating patients with chronic pain and other chronic disorders for implanted devices and their expectations, she is cleared for the trial. An appeal for approval for SCS is required from her insurance company, but it eventually is permitted and plans proceed for scheduling the trial procedure.

Variation: *The patient refuses the preprocedural psychiatric evaluation* – The psychological assessment of a patient before considering SCS is necessary to determine the readiness and appropriateness of this therapy for a patient. This evaluation can be broken down into three key components: (1) the cognitive capacity for mastering device technology, (2) the motivation for personal improvement and collaborative effort that a given patient has for working with the pain management staff, (3) identification and improvement of psychopathology that may preclude a good outcome from the SCS therapy. In many cases, ensuring patient education of the SCS procedure and technology encompasses most of the visit with the psychologist. The trial procedure can be much more productive when the patient is mentally prepared for the temporary trial phase. Identifying supportive family members who can assist the patient with the technology and offer support can enhance the chances for success.

<u>Variation</u>: *The patient is on chronic rivaroxaban because of a history of pulmonary embolus after deep vein thrombosis* – Guidelines put forth by the North American Neuromodulation Society and American Society of Regional Anesthesia and Pain Medicine require cessation of all anticoagulant medications before a percutaneous trial of SCS. The manipulation of leads in the epidural space during the placement is regarded as a high-risk procedure for intraspinal bleeding. In this regard, anticoagulants are held for the recommended times, in this case, 3 days before the planned procedure. Anticoagulation therapy is held for the duration of the trial, and the plan for cessation of anticoagulants is communicated and agreed upon with the prescribing physician. After lead removal, anticoagulation is resumed 24 hours later.

Scenario (continued)

The patient receives instructions for chlorhexidine washes the night before the procedure as well as the morning of the planned SCS trial. After consent and intravenous access followed by anxiolysis, the patient is brought to our fluoroscopy suite. Antibiotic prophylaxis is given using 2 g of cefazolin. An allergy or a history of methicillin-resistant *staphylococcus aureus* often necessitates a switch to clindamycin or vancomycin. The patient is placed in the prone position. Pillows placed under the abdomen to reduce lumbar lordosis to open the interlaminar space and facilitate lead placement in the upper lumbar or low thoracic levels (T11-L3). For upper extremity, access is usually in the upper thoracic levels (T1-T4) with pillows under the chest and a prone pillow to support and keep the neck midline. Depending on the patient's levels of anxiety and/or pain tolerance, light or moderate sedation is used to apply local anesthetic for the epidural needle tracts. Epidural access is gained approximately 1−2 levels below the targeted access vertebral level, which in this case was T11-12. The 14-gauge needle is directed in a shallow needle trajectory using a paramedian approach to the targeted epidural space for the low thoracic-high lumbar region. The upper thoracic levels have much smaller interlaminar spaces necessitating a steeper needle insertion angle, with only one vertebral level below and in a paramedian fashion. Anteroposterior, contralateral, and lateral views are used to guide the access needles and cylindrical leads to target levels. The stimulation is then activated, and the paresthesia is set to optimize coverage by real-time patient feedback. It is important to keep the patient lightly sedated to ensure the adequate coverage of patient pain. She tolerates the procedure well and has good verbal confirmation. Her foot and ankle area is covered by paresthesias during the placement. The lead is then secured on the skin with a strain-relief loop covered by a Tegaderm but leaving the very end of the lead accessible to connect to the trial stimulation device.

> **Variation**: *The patient refuses to have more than one procedure* – In some centers in the US and, in particular, in Europe and Canada, a permanent or lead-anchored trial is used initially to determine the efficacy of SCS. The obvious advantage is that the leads' position and programming on trial are identical to the final implant. There is no procedural variation, and the radiation exposure is reduced as well as the most challenging part of the procedure which is lead placement is complete, thereby reducing the risks of repeating the procedure. In our own center, permanent trials are very uncommon because of the risk of false-positives and wound complications. An anchored trial commits the patient more to the procedure and the industry representative may take a stronger interest in the patient's case than usual.

> **Variation**: *The pain is primarily isolated in the foot only* – The wider the distribution of the pain in the lower extremity, the easier it often is for SCS paresthesia-based systems to ensure coverage and have the greatest potential for efficacy. Lead position is often best low from T11-T12/L1 disc level to capture the foot. But at times, even with this strategy, the stimulation can be positional, with inadequate coverage, or be associated with extraneous stimulation. This likely reflects the transitional nature of the spinal cord to cauda equina, with additional cerebrospinal fluid making selective stimulation of large fibers to the foot challenging. High-frequency, non–paresthesia-based SCS has now been reported to offer relief of pain in the foot of patients with CRPS in our own experience. This approach will need further validation by future investigations.

> **Variation**: *Pain has recently spread to the upper limb* – The spread of pain to involve other extremities is not uncommon in CRPS cases. The mechanism is felt to relate to neurogenic inflammation with microglia and astrocyte activation (17). Proliferation of CRPS to involve additional appendages often lacks physical findings. Clinically, spread in CRPS is usually to the contralateral or ipsilateral extremity. SCS

therapy for CRPS affecting the upper extremity has the advantage of the option to stimulate the spinal cord or exiting nerve roots/dorsal root entry zone. The disadvantage is the positional nature of stimulation, particularly with cylindrical leads. The only means to reduce positional stimulation with percutaneous leads programmed to deliver traditional paresthesia is far lateral epidural placement of the leads to stimulate the cervical nerve roots. This permits leads to lock in the "gutter" and stabilize the position onto the nerve roots. Additional current options are to convert to subthreshold or high-frequency modes of stimulation. Future closed-loop systems may provide more consistent stimulation as well.

Scenario (continued)

The patient spent the next several days resting and engaging in only minimal activities. The needle access area in her back was somewhat uncomfortable, and initially, she had too much distraction from this pain to assess the trial benefit. Eventually, on day 4, she began to realize there was much less allodynia and that clothing and sheets could touch her foot again without eliciting severe pain. She was also able to sleep better than she did in years. She returned to have a small amount of reprogramming from the representative to try to eliminate some stimulation from her upper leg, which was able to be accomplished. After 7 days, she returned to the clinic where the lead was gently pulled out and a small dressing applied to the opening. There was no evidence of any infection, although there was some redness still near the entrance site. She was quite happy with the trial overall and wanted to schedule permanent implantation as soon as possible.

> **Variation**: *The lead became dislodged on day 3 after only half a day of benefit for the patient* – There is not a clear consensus of trial duration. The concern is mitigating the placebo effect and assessing the degree of improvement to gage long-term success of permanent SCS therapy. There are also the concurrent natural fluctuations of pain intensity observed in chronic pain states. Standard trials in Europe may last up to 3 weeks to mitigate placebo effect, but this approach is felt to enhance the risk of surgical site infection. A good deal of patients, even with a percutaneous trial, experience postprocedure pain at the first day and necessitate an additional day to adjust to the stimulation sensation. Therefore, a temporary trial for only 3 days provides the patient at most 1 day to assess the device efficacy. Keeping in mind the nature of chronic pain and placebo effects, there is a strong chance of a false-positive or an indeterminate result. This approach runs the risk of inappropriately recommending a permanent implant. Once a patient reports substantial improvement in pain intensity and function, either with traditional stimulation or other modes such a high frequency, it is

important to extend the trial to ensure that the relief is consistent. Patients who have true benefit from the trial appear with more vitality and seem much more comfortable. It is not uncommon for them not to want the temporary lead to be removed and inquire how quickly they can receive the permanent implant. Those patients who seem frustrated and seem to be looking for relief need direction and support so that they do not proceed onto surgical insertion of the device. Temporary trial durations that range from 5 to 7 days are more commonly used in clinical practice.

<u>Variation</u>: *The patient loses relief during the* temporary *trial* – Unlike anchored trials, temporary trials involve stabilizing the leads onto the skin. While leads can be sutured to the skin, they are more secured with biooocclusive dressings and medical tape. There is no proven superior method of temporary lead fixation. Patient instructions and cooperation are central to ensuring that lead positioning remains stable. Leads can move at any point during the trial and result in a loss of relief or no relief at all. In paresthesia-based systems, the patient no longer has topographical coverage, and this can quickly be determined. The patient may return to the clinic and have the device reprogrammed. During a trial of high-frequency SCS, securing the leads and ensuring patients follow instructions is of higher importance because of lag time for analgesic onset. A loss of analgesia is most often a lead migration on trial but does at times require verification by fluoroscopy. Imaging will allow for more rapid determination of the appropriate bipole in the lead to be programmed. Very commonly, leads can be reprogrammed to recapture pain relief. This does require a patient to come back into clinic, and the trial may need to be extended to reestablish therapy and verify pain relief. The restoration of analgesia does help support the validity of the high-frequency SCS trial. Some practitioners obtain final fluoroscopy imaging before lead pull to confirm the location of the leads and the exact location of the bipole at the end of trial. In cases where the leads have nearly pulled out or are obviously not in a position that have any chance of providing therapy, the trial must be redone. On account of SCS being an excellent last resort option for improvement in quality of life, a well-conducted trial of SCS is of central importance to our patients.

Chapter 3

Percutaneous spinal cord stimulation trial for cervical lead placement

Mohamed Osman, Musa Moris Aner
Anesthesiology, Critical Care, and Pain Medicine, Beth Israel Deaconess Medical Center, Boston, MA, United States

Scenario

A 55-year-old male presents in the clinic with chronic right upper extremity pain. He developed complex regional pain syndrome/causalgia after trauma to the right forearm in the several months after open reduction internal fixation. Ultimately, he came to have tried multiple neuropathic medications, multiple stellate ganglion blocks, physical therapy, acupuncture, and cognitive behavioral therapy in attempts to treat this disabling condition. However, he also had been unable to continue working in a physical warehouse job or in their administrative area at a desk job because of the significant pain with any activity. After further discussion regarding the limited options for him, the idea of a spinal cord stimulator in the cervical spine was raised. He expressed interest in trying almost anything and read literature and looked at websites provided to him. After another visit in clinic to discuss other aspects of the procedure, risks, benefits, and long-term aspects of potential programming, implantable pulse generator site problems, and potential tolerance of stimulation, he decided to proceed with a percutaneous cervical spinal cord stimulation trial. We scheduled this for him within approximately 4 weeks after obtaining psychological clearance and insurance approval.

> **Variation**: *The patient is on anticoagulation because of prior history of a pulmonary embolus* – The concerns for management of anticoagulation in the perioperative period of percutaneous spinal cord stimulator trials are very similar to the concerns found regarding lumbar stimulator placements (see Chapter 1). The same principles apply. A close conversation with a patient's cardiologist, hematologist, and/or primary care physician regarding when to stop and resume

The Neuromodulation Casebook. https://doi.org/10.1016/B978-0-12-817002-1.00003-1
21

anticoagulation as well as the safety of the process needs to be performed and documented in detail prior to proceeding with the procedure. At the current time, American Society of Regional Anesthesia
guidelines on anticoagulation are the standard of care in use. In general,
however, most procedures require all anticoagulation be stopped
5–7 days prior to the procedure and for a minimum of 3 days after. In
some more critical cases, a patient may need to be bridged with lovenox
to the procedure as it can be eliminated only 6 hours prior to the
intervention safely. However, after the procedure, a minimum of 3 days
without anticoagulation would still be required. If a patient is deemed
to be unsafe to be off of anticoagulation for that long, then the procedure (being elective) would need to wait until the patient is able to
meet these requirements.

Variation: *There is a prior history of anterior and posterior cervical
fusion at C5 through C7* – It is important to confirm the levels affected
in prior surgeries as well as the detailed findings in the operative reports
from prior cervical fusions. Of most critical importance in placing
cervical leads is whether posterior bony and ligamentous elements have
been removed in previous operations. A significant number of patients
with cervical degenerative disc disease do not undergo a posterior
laminectomy but solely a posterolateral fixation and fusion. Computed
tomography scans are most helpful in determining how much bony
anatomy was left in place. If the laminae are intact, a percutaneous
approach could be viable. Sometimes, in the presence of a unilateral
hemilaminectomy, a trial lead can still be inserted by targeting the
intact lamina site and threading a lead past the previous surgical field
into position.

Variation: *There is a history of severe rheumatoid arthritis with
limited range of motion of the neck* – Positioning is of utmost
importance in cervical placement of spinal cord stimulator leads.
Typical entry is at the upper thoracic levels at T2-T3, through T4-T5
interlaminar entry points. It is preferable that there are at least two
vertebral segments below the lowest electrode on the lead. The neck
should remain as neutral as possible, with a slight forward flexion, that
is supported by one pillow under the patient's chest and a gel pad under
the forehead. Extremes of flexion and extension should be avoided for
ease of epidural entry as well as threading and steering the lead.

Variation: *The patient cannot tolerate the arms "tucked in" position
during the procedure because of severe allodynia and hyperalgesia of
the right upper extremity* – Upper extremities are ideally placed in a
"tucked in" position during the procedure. However, in a patient with

complex regional pain syndrome I or II, this position could be quite uncomfortable to endure for the duration of the trial lead placement. Upper extremities could be placed on arm rests with adequate padding. A contralateral oblique view used during epidural access could be useful in getting the shoulders out of the visualized field.

Scenario (continued)

This gentleman is brought into our procedure room on the day scheduled, having washed previously with Hibiclens and having been nil per os since midnight the evening prior. He is slightly anxious but mostly looking forward to seeing whether this therapy can help him. He is positioned on the operating room table after getting mild sedation and adjusting rolls and pillows to accommodate his head and neck positioning to achieve a relatively neutral alignment of the cervical spine. While maintaining verbal interaction through the procedure, he is given local anesthetic in the region of needle access in the upper thoracic spinal area. A slightly paramedian approach is taken with the placement of the 14g needle using typical loss-of-resistance technique. The electrode is advanced into the epidural space after removal of the needle stylet. It is gently advanced under fluoroimaging using multiple fluoro angles and views to make sure the lead is continuing on the dorsal margin and not anterior. Initially, the lead will not readily align on the right side of the anatomic midline. The location is desired in this case because of his dominant pain in the right upper extremity. After several attempts, an eight contact lead is left in roughly the midline area where testing showed mostly left-sided stimulation, although some contacts gave bilateral stimuli. A second lead was then placed and able to be steered to the right side of the lead placed previously, using the previous lead as a boundary to prevent the lead from straying to the midline and left. This worked well and allowed for the second lead to yield primarily right-sided stimulation. Both leads were left in place and secured with a strain-relief loop on the skin with Tegaderm. The ends were attached to the stimulation device and once clear of the sedatives, he was sent home with several programs to try during the trial. Coverage appeared to be somewhat positional as it often is with cervical leads, but it was able to cover about 90% of his painful areas in the right upper extremity.

> **Variation**: *The patient has severe spondylosis of the cervical thoracic junction with poor identification of landmarks, including pedicles and interlaminar space* – Cervical spine spinal cord stimulator placements are significantly impeded by positioning of the patient as well as anatomic variation. Identification of certain landmarks such as pedicles and lamina structures is key to successful placement. Scoliosis and/or torticollis can impede positioning and identification of anatomical midline in these patients. Pattern recognition is of key value in

determining bilateral pedicles, spinous processes, as well as laminae. A slight cervical tilt is typically used to open up interlaminar spaces as well as aligning vertebral endplates. Once parallax is corrected, the straight line joining left and right pedicles typically marks the lower laminar edge of the interlaminar opening. Once positioning is confirmed, a slight reverse-Trendelenburg correction of the table can help with operator's access to the upper thoracic segments in a more ergonomic approach.

Variation: *There is unclear epidural access at T2-T3 space with loss of resistance using preservative-free saline* – The cervicothoracic junction is well known to have false-positive loss of resistance because of failure of completely fused ligamentum flavum in the segments. Significant attention needs to be patent to the angle of entry, along with fluoroscopic guidance to confirm proper epidural access. Repeat confirmation using loss of resistance to air, as well as, injection of a very small amount of radiopaque contrast could be used to confirm the epidural space. The lead should not be advanced unless there is clear confirmation of epidural access.

Variation: *The lead is difficult to get in position and causes paresthesias with every advancement* – Epidural access in a very acute angle can cause irritation of the dura in introduction of the spinal cord stimulator lead. Axis needs to be as flat as possible to the epidural space. Sometimes, a reinsertion of the epidural needle 1-1½ vertebral body spaces lower helps the access angle just enough to advance the lead smoothly. Additionally, if the lead advances and causes more of a radicular type of paresthesia, particularly related to the level of the lead at the time, then the lead is probably too lateral and an anteroposterior view may help confirm this.

Variation: *A cerebrospinal fluid (CSF) leak is noticed upon epidural access with the Tuohy needle* – A dural puncture at the level of the cervicothoracic junction needs to be clearly identified as the risk of injuring the cord with either the epidural needle or a spinal cord stimulator lead is of much higher risk at these segments than at the lumbar spine levels. If there is a dural puncture, the decision can be made either to abort the procedure or proceed at a level above the current insertion site. Given the limited access site in the upper thoracic segments, proceeding with a higher epidural access might not be feasible. Often, the patient would likely benefit from a rescheduled procedure at a later date. Should the appreciation of CSF leakage occur when the lead is being advanced, the lead should be removed immediately. Checking impedances (which would be abnormally low if the lead is within the CSF) or getting stimulation with extremely small amplitudes of stimulation all indicate that the lead is likely subdural and that there can be other means of validating this suspicion. Consideration should be given for placing a blood patch in either scenario with a CSF leak. It may be easiest to accomplish right at the moment because the patient is there in a setting where such a procedure can be readily performed. However, there is of course risk for cord compression as well and additional percutaneous access which slightly raises the risk of infection, so all aspects of risk benefit should be weighed.

Variation: *The patient complains of painful stimulation at his chest wall and his ribs* – This might be an indication of a "gutter" or too far lateral or ventral epidural placement. An anterior–posterior and lateral fluoroscopic image should be used to verify placement. The lead likely needs to be repositioned to a more midline and posterior epidural space. Although unusual for the lead to be in the mid to upper cervical spine, it is still possible to see this kind of variation. Also, positioning complaints should be assessed at the time in case the patient is experiencing pains related to how they are lying on the gel rolls and other cushions during the procedure. Finally, patient electrocardiogram and other vital signs may suggest there is a cardiac etiology for such pain, although one would not expect it to be then related to stimulation being applied or turned off.

Chapter 4

Percutaneous permanent procedures

Tom Simopoulos, Cyrus Yazdi

Department of Anesthesiology, Critical Care Medicine and Pain Medicine, Division of Pain Medicine, Beth Israel Deaconess Medical Center, Boston, MA, United States

Scenario

A 33-year-old female in otherwise good health suffers from chronic left cervical radiculopathy despite adequate decompression and fusion at the C5-6 level. She underwent a temporary percutaneous trial of spinal cord stimulation (SCS) and derived excellent relief with at least an average of 75% pain intensity reduction in a 7-day trial. She strongly wishes to go forward with the permanent implant. The permanent implant is scheduled for the procedure center in the next few weeks after further clarification of questions she had as to risks and details of subsequent pain and programming.

> <u>Variation</u>: *The trial procedure was very painful and the prone position not well tolerated* – Patients with chronic low back and/or neck pain may tolerate the prone position for only a limited duration on account of exacerbation of symptoms. It may be so brief that they cannot reliably get through the trial or permanent percutaneous placement adequately. The case may be further complicated by high levels of patient anxiety and complex pharmaceuticals, in particular opioids. The balance of sedation depth and patient comfort during monitored anesthesia care can quickly become in conflict with one another. This can be further complicated if there is coexisting obesity with a challenging airway that favors easy obstruction resulting in rapid hypoxic consequences. For safety reasons, a case with the aforementioned complicating factors may be referred for surgical paddle placement.

> <u>Variation</u>: *A second lead is needed for adequate coverage but cannot be placed in the permanent procedure* – There may be technical challenges in placing more than one percutaneous lead. Even though the trial may have had one or even two leads placed, the replacement of

The Neuromodulation Casebook. https://doi.org/10.1016/B978-0-12-817002-1.00004-3
27

leads in the permanent procedure can be hampered by adhesions or other anatomy and a second lead becomes needed but cannot be positioned well enough. To ensure adequate long-term therapy, in many cases, more than one lead is necessary. While a paddle insertion may be more invasive, the complex arrays that it offers can permit durable results and reduce the need for future revisions. The main issue for the surgeon will be postoperative pain management as these leads are often able to be placed under general anesthetics now. A preprocedure plan with the anesthesiologists in the preprocedure testing center or with the referring pain physician can help provide an improved postoperative patient experience. If these options were not easily available, then a postoperative consultation with the acute pain management service will help enhance the comfort in these challenging patients.

Variation: *Stimulation is too intermittent and varies extensively with neck position* – When considering a trial of traditional SCS in the cervical region, it is imperative to assess for positionality of stimulation intensity during the trial. Despite adequate paresthesia coverage, if the stimulation is too erratic, it becomes impractical. This is especially true if percutaneous leads are placed in the midline epidural space, rather than in the paraforaminal region which can serve to stabilize in the narrower aspect of the epidural space so as to better couple the lead with the neural stimulation targets. This strategy should be adopted during the temporary trial, and an assessment of the degree of positional stimulation should then follow. If stimulation has too much variation with neck position even after doing this, then the patient can be referred for a paddle electrode. Paddle leads are typically more stable in the long term and sit slightly closer to the cord overall because of their geometry in the epidural space. In today's era of multiple waveforms, non-paresthesia modes of stimulation can be used to reduce this limitation and variability of paresthesia intensity.

Scenario (continued)

The woman proceeds through preoperative evaluation and prophylactic assessment for the permanent implant procedure. Infection prevention is important. The complication of infection is the most common biological complication and is estimated to range between 2% and 3%. Therefore, most of the evaluation and assessment of the patient is focused on mitigating this problem. The Neurostimulation Appropriateness Consensus Committee (NACC) has put forth key recommendations to reduce surgical site infections. The modifiable surgical site infection risk factors identified include diabetes, smoking, corticosteroids, and even high-dose chronic opioid therapy. While these factors have not so far been specifically associated with higher risk of infection after SCS implantation surgery, NACC does recommend specific strategies. Smoking cessation for at least 4 weeks prior to implantation surgery or application of a nicotine patch is recommended. HbA1C optimization and reduction of high-dose opioids should be performed prior to SCS procedures if possible. Down titration of opioids prior to surgery is thought to reduce the immunosuppression caused by these medications, and at the same time, it may lessen the difficulties of postoperative pain management.

Preoperative screening for methicillin-sensitive and methicillin-resistant *S. aureus* is currently recommended to minimize the risk of surgical site infection. Decolonization of *S. aureus* prior to surgery may reduce infection rates close to 60% (4). NACC endorses the use of routine preoperative nasal swabs and decolonization with mupirocin and chlorhexidine baths. Practically, it can be challenging to meet all preoperative optimization steps, but moving toward clinical pathways that continue to incorporate and ensure infectious mitigating steps helps reduce patient morbidity and reduce health-care costs.

The patient receives instructions for chlorhexidine washes the night before the procedure as well as at the morning of the planned SCS trial. After obtaining consent and intravenous access followed by anxiolysis, the patient is brought to our fluoroscopy suite. Antibiotic prophylaxis of 2 g of cefazolin is given. In an identical fashion to placement of the trial leads, pillows are placed under the abdomen to reduce lumbar lordosis to open the interlaminar spaces and facilitate lead placement in the upper lumbar or low thoracic levels (T11-L3). Specifically, in this case, access is usually in the upper thoracic levels (T1-T4) with pillows under the chest and a prone pillow to support and keep the neck midline. Positioning is very important because of the light sedation during lead placement and the extended time that may be necessary to complete the procedure.

Fluoroscopy is used to guide 14-gauge epidural needles via a paramedian approach to the targeted interlaminar space usually T1/2 starting skin entry one vertebral body below. Anteroposterior, contralateral, and lateral views are used to access the epidural space and advance leads. At least several electrode contacts are placed above the target level of stimulation on account of lead migration of one contact which is common in the cervical region even in experienced hands using mechanical anchors. As the lead is advanced for her, she has some exquisitely sharp pain in the neck, possibly from dural irritation or ligament irritation. The lead is backed down and readvanced, and it is avoided.

Two leads are placed into proper alignment and level (confirmed by the awake testing), sedation is enhanced, and local anesthesia is given. A midline 1- to 2-inch vertical incision is made at the level of the needle insertion, and dissection is carried out down to the supraspinous ligament. The subcutaneous tissue is dissected by electrocautery, and the adipose tissue surrounding the epidural needles is removed to clearly expose their trajectory to the fascia. A small defect in the fascia is created right where the epidural needle pierces it, so as to allow the nozzle of the anchor to pass through the fascia. This is an important step to prevent leads from buckling back and kinking. This maneuver has the benefit of limiting caudad migration and enhancing lead longevity. Nonabsorbable braided sutures (0 silk, 0 ethibond, or 2 fiberwire) are used to link the anchoring device to the supraspinous ligament in most cases. The majority of lead anchors now use a mechanical lock to link the anchor to the cylindrical lead. This modification in the anchoring mechanism has substantially improved lead migration and need for surgical revision. With the anchor secured, attention is turned to the implantable pulse generator (IPG) pocket and tunneling.

There are several important considerations for success with implantation of the IPG. The first is to ensure that the patient will be comfortable with where it is inserted because of hardware-related pain, especially in chronic pain patients. The usual choices are buttocks, flank, and abdomen. Discussion had taken place in the preprocedure holding area and marked beforehand in the sitting position to best locate the pocket as the soft tissues naturally set. This is of critical importance because these locations can appear very different once the patient is in the prone position. The infraclavicular region is less commonly used in SCS practice but can be an option for cervical leads.

In this case, a flank location is chosen. A 2-inch incision is made at least two finger breadths below the scapular region, and a small subcutaneous pocket is created just below Scarpa's fascia with blunt dissection and limited electrocautery use. Because of the need to recharge many of these devices, it is important to keep depth in mind and limit dissection to 1.5–2 cm from skin. The leads are tunneled into the IPG pocket using the tunneling tool provided in the kit. The pocket is irrigated with saline, and the IPG is placed into the pocket after connections are made and impedances are checked. While flipping of the IPG is very uncommon, it can occur with abdominal placement. Patients who are overweight with poorly developed musculature can develop enough tissue laxity where the IPG can flip, making recharging a challenge. All incision lengths and dissection should be kept to the smallest size as well as avoiding excessive electrocauterization of tissue are useful techniques to reduce risks of infections. Nonabsorbable monofilamentous sutures can be used to secure the generator to the fascia. A multilayer closure is used to encourage rapid healing, consisting of 2-0 vicryl for the fascia, 3-0 monocryl for the deep dermis, and 4-0 monocryl for the subcuticular tissue. Finally, a skin adhesive (dermabond) is applied to seal the incision with bio-occlusive dressing. A multilayer closure in practice promotes rapid wound healing to reduce the risk of dehiscence.

The IPG pocket site accounts for the majority of infections, resulting in a number of infection-mitigating strategies being purported over the years. The NACC has published the following key intraoperative recommendations to reduce infectious complications: timely antibiotics, chlorhexidine/isopropyl alcohol skin preparation, double gloving, C-arm drapes, iodophor-impregnated drapes, and minimizing operating room traffic. Future modalities to reduce infection rates are vancomycin powder and impregnated antimicrobial envelopes placed into the wounds.

Two years later, the patient has developed failed back surgery, necessitating a permanent SCS device after a prior successful trial. She has a fusion from L4/5 to L2/3 level with superior adjacent segmental stenosis. She has persistent painful lumbar radiculopathy in an L5 distribution on the left-hand side. The patient has again failed to find meaningful relief from physical therapy, multimodal analgesia, and epidural injections. In clinical practice, it is not uncommon to encounter patients who necessitate spinal cord stimulators to manage both cervical as well as lumbar radiculopathies. These individuals may represent a subset of patients who are prone to the development of neuropathic pain states.

A trial can be easily undertaken in this case as well as a permanent percutaneous implant by simply accessing the above levels. The issue to keep in mind is that the stenosis at L1/2 will likely need future decompression. Placement of leads in this region with strain relief loops in the neighboring tissue may be in the way of future surgery and run the risk of damage during the decompression operation. In this case with high lumbar level pathology such as stenosis or disc extrusion, a thoracic level placement is likely more prudent with an open surgical paddle placement. Programming and recharging can differ significantly between the cervical and thoracic levels, so having two separate IPGs has, in our experience, worked well for patients.

Chronic pain patients necessitate specific discharge instructions with emphasis on symptoms and signs of infection. Despite current retrospective data suggesting prolonged antibiotic use to 5 days may reduce future surgical site infection, current recommendations by NACC do not recommend use of antibiotics beyond 24 hours. Most of the data on antibiotic prophylaxis are regarding preincision dose only, which is optimized with respect to patient weight and time to incision. Bio-occlusive dressing for up to 48 has been associated with risk reduction.

Variation: *The patient intends on a future pregnancy and asks about use of the device* – Despite very small current densities delivered right to the spinal cord, the electromagnetic radiation given off by the device is of uncertain safety to the fetus. Similarly, the consequences of poorly controlled chronic pain with its negative impact on mood are of unclear effect on the developing fetus as well. Although there have been case reports reporting the variable use of SCS during pregnancy, the current clinical recommendation is to keep the device in the off mode. Thinking forward, it is also recommended to keep the pulse generator away from the abdomen in patients who may become pregnant in the future.

<u>Variation</u>: *The patient reports driving a car with the SCS device generating active paresthesia* – Most patients who derive good relief from SCS report using the device in the on mode while driving. It is common for neck, back, and associated radiating extremity pain to intensify with prolonged driving trips. With the exception of a high frequency of 10 kHz, current manufacturers do not recommend driving while using SCS. Therefore, we refrain from endorsing use of the SCS devices while driving and recommend following the manufacturer's guidelines.

Scenario (continued)

The patient presents 2 months later with complains of pain at the IPG site. The immediate concern is to exclude infection by clinical examination and if necessary assess complete blood count, sedimentation rate, and c-reactive protein levels. Once infection is excluded, hardware pain is not uncommon in chronic pain patients and can be significant enough to necessitate repositioning or even explantation. Most cases improve in the ensuing months. While no specific pharmacological approach is highly effective, topical lidocaine or diclofenac as a patch can help in some cases. Anti-inflammatories can also help. Opioid medications are usually limited to just the first several weeks and should be weaned rapidly. In rare cases, the IPG needs to be revised more deeply or to another location.

<u>Variation</u>: *The patient returns after 4 months after the insertion of the thoracic SCS and complains of pain just below the mid thoracic spine that disappears only when the device is turned off entirely* – The pain is directly over the spinous process at the T9-10 level. The suspicion in this case is ligamentum flavum stimulation (LFS). On evaluation, it is important to exclude wound-related pain and spinal pathology such as compression fracture. The midspine pain correlates with the position of the stimulating portion of the lead(s). LFS cannot be overcome through reprogramming and results in progressive loss of use of the device. This is observed although on rare occasion only with cylindrical leads because of the 360 degrees of stimulation pattern. In the past, the only way to resolve this was to replace the lead with a paddle type lead, but this may now ameliorate by the application of subthreshold stimulation.

Chapter 5

Paddle placement for failed back surgery syndrome

Jeffrey E. Arle[1,2]

[1]*Neurosurgery, Harvard Medical School, Boston, MA, United States;* [2]*Neurosurgery, Beth Israel Deaconess Medical Center, Boston, MA, United States*

Scenario

A 42-year-old woman is referred for placement of a paddle lead and implantable pulse generator (IPG) after a successful trial by an outside pain physician in the community. The woman has had three prior lumbar spine surgeries: initially, an L5-S1 R microdiscectomy, moderately successful, followed several years later by a generally unsuccessful L L45 microdiscectomy in which she had residual significant lower back pain but also numbness and some persisting pain in the L5 distribution of the L leg with any activity. This was then followed by a 2-level lumbar fusion using transforamenal lumbar interbody fusion (TLIF) approaches at both L45 and L5-S1 as well as pedicle screw fixation at L4, L5, and S1. She has had persisting significant lumbar pain after this surgery as well, now across both sides of the lower back equally with some pain in both legs, left > right.

> **Variation**: *Only one prior surgery* – Although the literature strongly continues to support considering spinal cord stimulation instead of reoperating, it also remains clear that if the patient has an obvious structural component to their continued pain and it can be readily addressed by a further surgery, then that surgery should still be performed first before stimulation. Such obvious structural components to one's pain might be a new herniated disk, broken hardware with a progressed listhesis, or severe stenosis above the level of a prior fusion. Decisions become trickier when the presumed structural component is less overt: fibrosis around a nerve root, foraminal narrowing, and facet arthropathy, possible pseudoarthrosis. Even with only a single prior surgery, if the patient has a less overt structural problem, reoperation is rarely indicated versus spinal cord stimulation, and the likelihood of a better outcome favors spinal cord stimulation.

The Neuromodulation Casebook. https://doi.org/10.1016/B978-0-12-817002-1.00005-5

Variation: *The patient has only back pain* – While it is common to have predominantly back pain in these cases of failed back surgery syndrome "FBSS," traditional spinal cord stimulation has been less successful than if leg pain is the predominant feature. However, with more recent developments (higher frequencies, burst patterns, sub-threshold stimuli, and more sophisticated traditional systems), getting back pain relief is more reliable in the long term (i.e., more than 1−2 years) than it used to be. Often pain relief can only be achieved to the upper buttock region, even with a "successful" trial where low back regions were helped. While I still feel compelled to let the patient know ahead of time that managing their low-back pain is certainly possible and SCS is worth trying, it is often harder to obtain with programming and hold on to, although technology is continuing to improve in this area.

Variation: *There have been no prior surgeries yet, just worsening back and leg pains* – This situation is a fascinating area for progress and development in spinal cord stimulation. While some components of the patient's pain in such a case can be neuropathic in origin, it is almost inevitably a mixed neuropathic/nociceptive picture. SCS can still be very effective in such cases, and with third party payors increasingly denying approval for what they see as nonoperative back and leg pain, patients who fail medical management and other less invasive therapies (e.g., physical therapy (PT), epidural steroid injection (ESI's), and radiofrequency ablation (RFA)) may be left with little else to try, other than SCS. The ability to perform *a trial* with SCS always makes this therapy an option to consider.

Scenario (continued)

She failed two courses of physical therapy, medication use including gaba-pentin to 2400 mg/day, muscle relaxants (Robaxin 750 mg bid), and ibuprofen 800 mg up to 4x/day, as well as two oral steroid tapers, five epidural steroid injections, two medial branch blocks, and a radio frequency ablation at L L45 and L L5-S1 and R L45. She has tried chiropractic care for 6 months with only occasional benefit that is not long-lasting, a TENS unit, a heating pad, and oxycodone, which she now takes 10 mg up to 3x/day. Her pain is typically eight or nine on a 10-point visual analog scale. The oxycodone reduces it to perhaps a six she says, on a good day.

Variation: *The patient had no prior injections* – This patient has a classic picture that readily provides a solid indication for trialing an SCS. Had she not had any steroid injections, epidural or medial branch blocks, she would still be a candidate for SCS in my view, but she may have more peace of mind moving forward with it if she knew she had also failed these injections. Success for the long term from a few injections in this kind of case is extremely rare, and there are risks, albeit very low. I would typically suggest her to try one or two, just to know, but if she adamantly refused to do them for some reason, I would still move forward to an SCS trial.

Variation: *The patient had not tried muscle relaxants or any gabapentin* – Often a large component of FBSS is muscle spasm, widespread or focal, persisting and/or episodic. Not having tried a muscle relaxant is a mistake that is easily corrected. Many patients will need to stay on one for the foreseeable future even after the stimulator is placed. I see patients commonly, as well, who have been given one muscle relaxant, felt "spacey" or "loopy" or overly tired and then dropped it completely and not tried any of the four or five others available. Their primary care physician (PCP) or pain physicians should be able to run through these in a month or two to really see that none of them help before giving up on them. They are not addictive, and no one should have a problem prescribing them. Gabapentin is also helpful for neuropathic pain in many cases, although in my experience, it is rarely potent enough to allow patients to stay away from seeking further input and treatment with FBSS. Part of this is often because they are severely underdosed with gabapentin. The standard dose that needs to be tried, ramping up over about 3 weeks, is 1800 mg/day, typically split into three 600 mg doses. Many patients are given 100 mg/day or 300 mg doses only to take at night. These doses are wholly inadequate for the typical pain associated with FBSS. In most cases, both a muscle relaxant and adequate gabapentin should have at least been tried before moving to an SCS trial.

Variation: *The patient was taking 10 mg Dilaudid and 40 mg MS-Contin every day without benefit and asks if she can have her meds refilled before surgery* – Of course, physicians have different styles and preferences. I do not like to have any role in managing a patient's pain medication, narcotics or otherwise, prior to surgery. After surgery, I do manage their medications until a sufficient time has passed to be notably reduced toward a baseline following resolution of surgical pain. For an SCS paddle placement, this is typically a few weeks. In this scenario, I would make this clear to the patient. However, these preoperative high doses of narcotics are going to make postoperative management very difficult. I tell patients if they can wean their narcotic requirement some amount prior to surgery it will make pain control after surgery much easier. That said, most of the time such patients cannot wean their medications much at all going into surgery, and it is in my view unreasonable to expect much reduction either. I try to make it very clear to patients ahead of surgery that our ability to control the surgical pain after surgery is not going to be easy. Usually, they want surgery anyway, it is difficult, and we eventually get through it.

Variation: *The patient has actually more back pain than leg pain and a TENS unit worked well, but only used it a few times for 15′ at a time* – This is an interesting situation that I have encountered many times. Often, patients are told they should only use a TENS unit for 15′ or maybe 30′ per day. Perhaps they tried it in physical therapy. If the patient has benefit with it and particularly if the patient is on the thin side and also has predominantly low back pain as opposed to leg pain, then a TENS unit continuously, or nearly continuously, is a viable option to test before proceeding with an SCS trial.

Scenario (continued)

Her films are reviewed from just prior and after her past surgeries. They show reasonable hardware placement and sagittal balance, and none of the screws or interbody devices clearly abut or compress a nerve root. There are postsurgical changes in and around the areas of prior discectomy and decompression but no obvious reherniation, listhesis, or stenosis. I then proceeded to discuss the trial she had as it seems that the idea of treating her with spinal cord stimulation is a good one, perhaps even long overdue.

Variation: *The patient doesn't have any recent imaging of their spine surgeries with them* – If the patient has been already successfully trialed elsewhere, which is the typical scenario, then I plan to proceed with SCS placement, but I ask that they get me their films so I can just make sure they do not have an obvious structural problem that could be operated on appropriately. It is possible that another surgeon simply did not want to deal with reoperation or operation to begin with on someone they perceive to be a difficult chronic pain patient but that the patient actually *does* have a straightforward structural problem. This is worth paying attention to and addressing, if for no other reason than, if the patient has a large herniated disk, for example, the SCS likely will fail in the medium to long term no matter what happened in the trial.

Variation: *Patient clearly has mechanical back pain and a broken pedicle screw with advanced listhesis from previous imaging* – As noted in the response previously, there are occasions when a patient has a more overt structural problem, and it should be addressed with another surgery. This is likely one of those scenarios. And, as noted, it is entirely possible the surgeons who have evaluated this patient ahead of me have not wanted to take this on, even though it is indicated instead of SCS. One way to think about this is that an SCS is not likely to be successful unless this condition has been addressed surgically first.

Variation: *Patient tells you they have new pain in back and leg predominantly on one side probably in L4 distribution (prior work was L4-S1 levels) more recent than most recent spine imaging* – As with the other scenarios described previously in this chapter, new conditions and potentially overt structural problems need to be imaged appropriately and potentially addressed surgically ahead of SCS placement. On occasion, patients are trialed under these circumstances anyway and are sent to me for an SCS placement. Most likely, the pain physician managing the patient has already received opinions from surgeons that there is no surgery to be done, so they do not consider asking for my view on that. In this case, a new condition certainly warrants at least getting a new magnetic resonance imaging (MRI).

Scenario (continued)

She says she thinks it worked well during the trial and she is looking forward to getting this system put in as soon as possible. She wishes the trial system had not been removed as it was the only thing that has helped her over the previous 8 years. She says she would do the surgery today if we could, only half-kidding. On further questioning, it appears that the trial lead placement was a little challenging from what she could tell, as she was awake, but lying prone. She thinks that the doctor was having trouble getting the second lead in and positioned where he wanted it. She had a lot of pain in the first 2−3 days after the trial leads were put in, and she could not tell if there was any benefit. The representative from the company called her on day two and met her on day 3 to make some changes to the programs and get better coverage of her painful areas. Eventually on day 4, she thought she had about 50% reduction of back pain and 70% reduction of pain in her legs, though it did not cover the right leg very well. Her pain was worse on the left so she seemed satisfied with this.

Variation: *The trial was helpful only for a day, and then the leads became dislodged somehow and then were removed in the office* – A single day of trial wherein the stimulator was connected correctly, worked appropriately, and the patient felt benefit, is simply not adequate, and they need to be sent back for a second trial. It is disappointing for everyone involved, but it is the best way to handle this situation. For me, four or more days is ideal. Three days is marginal. Less than three should be retrialed.

Variation: *The patient says the system worked great for their legs but not so much for their back − they could feel it there but the pain was still there and their pain was also mostly in the back area to begin with* – It is imperative that one makes it clear that this patient is unlikely to be satisfied with *only* having success for their leg pain. If the patient did *not* feel the stimulation in their back, that is a different situation and reprogramming may ultimately solve that adequately. However, if the patient feels stimulation in their low back area, it is the correct area where their pain is located, and they do not get pain relief, then this is inevitably unsolvable, and it should be explained that way to the patient before they are implanted. They may decide that having the stimulator is no longer worth trying. It should be noted that this is the way to think about this situation if "traditional" stimulation is being used (i.e., stimulation where the patient feels the buzzy, vibratory like paresthesias of the stimulation). Nonparesthesia stimulation paradigms may still be successful, and these might be considered.

Variation: *You get the feeling that the trial was not really 50% better in any area — the patient may have secondary gain issues or is just desperate to have something done* – It is important that the patient be honest in their assessment of the trial. It is important that the physician realize that the patient may not be honest in their assessment of the trial. Patients in pain will often resort to trying anything to be out of pain, and if it means telling the surgeon they had 50% relief of their pain to be approved for permanent implantation many of them will say exactly that, even if it was less in actuality. Implanting these patients almost always fails in the long term. The placebo effect from stimulation is also between 20% and 30%, so a patient saying they had 30% benefit and wants it placed anyway is barely distinguishable from placebo. These patients typically fail in the long term. It gets difficult when a patient says they had well over 50% benefit in one area but less in other areas and they still want the permanent implant. I usually will move forward with surgery in these patients but it is important to make it clear that they may not get much more than that, if any, with the permanent implant.

Variation: *The patient says the system helped but most of their pains are mechanically-driven it seems and typically episodic in nature, short sharp pains sometimes here and sometimes there* – These types of pain patterns and character rarely respond well, especially in the long term, to SCS in my experience. I would strongly suggest to the patient that she is unlikely to have a great result in the long term despite what happened in the trial. If she had not been trialed yet, but was requesting a trial, I would dissuade her from thinking SCS would likely be a successful therapy.

Scenario (continued)

I then asked her how many days she had this benefit, whether it remained as good as she described during that time and whether this benefit was enough to go forward with the surgery, although I knew her answer to that from her prior comments. Interestingly, she said she had this benefit for about two and a half to 3 days but then the leads were pulled out in the office. I then spent several minutes going over a description of the surgery, the two incisions she will need, the typically larger amount of postoperative pain that she will have compared to the trial, that she will be asleep during this surgery, it takes about an hour, and that we do not need to wake her up in the middle of it. She says her spine surgeries were very painful and she cannot imagine anything will be as bad as those and that she is in pain all the time anyway so she is "used to it." I tell her that she probably will still be surprised at how painful it is even though the incisions are only a couple inches long, but that despite this pain, we will give her medication and try to make her comfortable and that most likely, about 95% of the time, my patients still go home the same day. I also discuss the possibility that the paddle lead we will place may not work as well as the trial, despite having been told that it should work better. And I discuss the statistics that favor getting better leg pain relief, especially in the long term, than back pain relief.

She was sent for a history and physical with her primary care physician who would then fax it to our office before the surgery. No labwork was needed as she was otherwise generally healthy. Prior imaging for her spine surgeries had imaged the region of the thoracic spine where the lead would be placed, indicating there was no significant disk herniation or stenosis there, but she also had no symptoms relatable to cord compression there or pain from that area. The typical preparation for surgery (hibiclens wash, not eating after midnight) was asked of her, and she was next seen in the holding area on the day of surgery.

Variation: *Currently taking Coumadin for atrial fibrillation* – These patients can undergo SCS placement, but they need to be off of their anticoagulation for 5–7 days ahead of surgery and for a minimum of 3 days after surgery. Consultation with their cardiologist or hematologist is needed to make sure that the patient can tolerate any additional risk if off their anticoagulation. Significant risk patients need to be bridged with a heparin analog up to surgery, but even then they need to be completely off anticoagulation after surgery for a minimum of 3 days.

Variation: *No previous thoracic spine imaging is available* – Thoracic MRI or postmyelogram computed tomography (CT) scans must be reviewed for SCS paddle lead placement prior to surgery. The surgeon must be aware of severe stenosis or disk protrusions ahead of lead placement. However, having stenosis or a herniated disk per se does not eliminate the ability to place a paddle lead. The surgeon may simply decompress more of the bone and ligament as they prepare access to the epidural space for lead placement. In the most extreme situation, the entire lamina may be removed and the lead sutured directly to the outer layer of dura with 4-0 Nurolon if necessary. In such cases, it is important to recall that the lead will not be as close to the spinal cord and the additional distance of cerebrospinal fluid will require more amplitude and may diffuse the precision of the field and make the stimulation more positional in nature.

Scenario (continued)

Typical IV access and plan for a prone surgery was made. Prophylactic antibiotics were given within 30′ of the incision (typically 2 gms IV Cefazolin unless there is an allergic concern). Also, we use general anesthesia with a total intravenous anesthetic technique (TIVA) and help orient the electrode using a stimulation technique through the electrode which generates electromyography (EMG) in the lower extremities. This can also be done in the cervical region and upper extremities. We monitor the spinal cord with motor evoked potential (MEPs) as well just to be able to check cord function if there are any concerns during the surgery. Fluoroscopy is also used to visualize the electrode and the spine in an AP view and determine the level. I have her sit forward on the stretcher to mark the incision for the upper buttock area based on whether she sleeps on one side or the other, where her belt line runs, if she has trouble using one arm or the other and so forth. This is where the IPG will be implanted, and I make mention of the fact that even in some cases, everything heals well and looks fine but the IPG may bother her, and in rare cases, we need to move it to another location.

She is brought in the room and put to sleep, intubated, and turned into the prone position with appropriate chest and iliac crest cushioning. Arms were placed at a right angle bend up and cushioned. The fluoroscopy unit is brought into place to determine the midline level of the thoracic incision. The incision is planned for a location just inferior to where the desired final location of the lead will be, based on the lead design, type of stimulation therapy (high frequency vs. traditional stimulation), and need to cover certain areas of pain (e.g., foot vs. low back).

The thoracic incision is made with a 10-blade, and dissection is taken down along both sides of the laminar margins using a monopolar cautery. Hemostasis is kept intact usually with the monopolar or bipolar as needed, and a cerebellar self-retaining retractor is used. Typically, about one and a half segments are visualized. The incision length is often about 3 cm, longer in larger patients. A fluoroscopic image is obtained with a marker in place angled superiorly to determine the level in the spine. This helos determine then what bone to remove. Rongeurs, the high-speed drill, and kerrisons are used to remove a portion of the spinous process and lamina, perhaps half of the laminar length, or some of the inferior lamina and superior lamina, depending on their sizes, the length of the lead, and the location. Hemostasis is obtained using the bipolar and bone wax as needed. Antibiotic irrigation is used. The dura is well visualized in this more or less rectangular access to the epidural space that has been prepared. The lead is then carefully inserted into the epidural space using a bayonet forceps, often holding it at the base of the flat part of the silicone. Ideally, the lead will sit midline or nearly midline by fluoro image, and the base of the lead will be visible in the upper edge of the opening. If the lead needs to be removed later when it is well-scarred in, this will be much easier than if the lead is slid more superiorly completely under the lamina.

Testing by stimulating through the lead using the connecting cables to the end of the lead and then off the field to the company representative to test the EMG generated at the four corners of the lead or other variations that help determine the orientation of the lead to the dorsal column of the spinal cord then proceeds. In a small number of cases, no EMG is able to be generated, and the lead is placed based solely on imaging. The lead is repositioned carefully until the position is satisfactory. Silicone anchors are sutured to the interspinous ligament with 2-0 silk, and after irrigation and a check of hemostasis, the retractor is removed, and the fascia is closed using 0 Vicryl.

At this point, the incision is made for the IPG site pocket. It is made in a slight curve, and with either blunt finger dissection or using curved Mayo scissors, the pocket is created to be just large enough for the IPG. It is packed with gauze, and the tunneler is then passed between the two incisions, the leads pulled through to the IPG site and connected to the IPG after drying them off, and the IPG placed into the pocket after removing the gauze, making sure there is hemostasis within the pocket. Often, the impedances are checked to make sure the connection is sound. Both incisions are then closed using 2-0 Vicryl and 3-0 running unlocked Nylon to close the skin. Dressings are placed, and she is turned back into the supine position, awakened, extubated, and taken to the recovery room for further care and start-up of the stimulator.

Most likely, she will be able to go home the same day after a couple of hours in the recovery room. The thoracic incision tends to be very painful, and in about 5% of patients, they need to be admitted for pain management.

Variation: *The lead is very difficult to get into the midline and each attempt leads to the tip deviating to the right mostly or to the left, at about the three-fourth implant distance* – This impediment, figuratively and literally, is the most common problem in lead placement. Some leads on the market are longer and require the lead to hold its position in the epidural space for a further distance, making them harder to place usually. In many cases, however, the lead will go right in and be reasonable. I have used, carefully, a 1-cm-wide malleable brain retractor blade as a blunt gentle instrument to clear some of the barriers in the epidural path. Kits usually include a plastic version of this instrument, but I have not found any of these variants available for this task useful. Sometimes, just diligence is needed to get the lead to maintain a midline position and alignment. Sometimes, however, it is necessary to move up one level and make a second (and even a third –

especially in the cervical spine) laminotomy to help guide the lead (with a nerve hook for example) and clear ligament or a midline keel from under the lamina in certain cases. Whether or not to do this extra surgery is a judgment call, but after many instances of this problem occurring, I have gotten a feel for when to abandon a single opening to place the lead and to make a second laminotomy. It ends up being some kind of combination of time management and how deviated the lead is each time it is attempted.

Variation: *Stimulation through the electrode is attempted but no EMG is elicited even at 20 mA* – I have had a handful of patients over the years of developing and using this technique in whom we cannot seem to obtain any EMG despite very large bipolar configurations involving the entire lead and very high amplitudes and pulse widths (20 mA and 500uS for example). While quite disconcerting, it is imperative that the Intraoperative physiology technicians provide strong assurance that it is not a technical issue with their equipment. In none of these cases has the patient been found to have a deficit postoperatively. It is unclear why we do not get EMG in some cases, but it may be related to the fact that the alpha motor neuron receives approximately 50 distinct sources of input and varying numbers of synapses from those sources. In some patients, maybe the number of synapses or their location, or both, from the dorsal column axons do not support these neurons reaching threshold adequately.

Scenario (continued)

She did well overall with the surgery and healing. I removed her sutures on day 14, and there were no concerns with the healing of the two incisions, only a slight tenderness at the IPG site. The company representative was there at the visit to spend some time on reprogramming. She reported that she was getting about 50% benefit in her back and 70%–80% benefit in her left leg and that it was missing some areas she wanted in her right leg. I reassured her that it was still early and that with time as the scar settled internally and some further reprogramming, it was likely to be better yet. I also reiterated that she should be as proactive as she wants to be in meeting and getting more programming. It may be three or 4 times at the beginning and then once a year or so for a couple years that are needed, or any variant thereof, and that she could always come in to obtain further tuning.

Variation: *She has some serosanguinous drainage from a corner of her IPG incision* – If it is truly serosanguinous and not pus, then it is reasonable to watch, perform dressing changes once or twice daily, and keep the patient on an antibiotic (cefazolin for example) until the wound heals. It usually heals and is not infected. I do not typically remove these devices unless they are clearly eroded or infected. If they need to be explored and there is not obvious pus found, I often will irrigate with antibiotic saline aggressively and place vancomycin powder around the wires or IPG and then reclose the wound and follow the patient closely. Often these devices can be salvaged under such circumstances, but not always. If there is obvious infection, the device (usually lead, wires, and IPG) all need to be removed. Waiting a few months and going back in to replace the system is certainly possible.

Variation: *She has significant "positionality" and cannot find a stable posture or position to get steady benefit, or the amplitude is too low or high* – This kind of problem occurs in a minority of patients and often can be programmed around or away to a large extent – but not always. In rare cases, it is too aggravating, and the patient is getting little or no appreciable benefit from the system. One potential solution to consider is placing a "dummy" lead made of pure silicone over the top of the current lead, in the epidural space. This must be performed carefully and with full knowledge of the cerebrospinal fluid (CSF) distance there – a CT can be helpful in assessing this. The cord and lead (despite artifact) can be more or less seen together and a decision made as to whether there is enough space to add a second lead volume, moving the current lead that much closer to the cord. If this can be safely accomplished, it will almost always eliminate or significantly lessen the positionality.

Variation: *Attempts to get better coverage lead typically to rib or abdominal stimulation, on one side or the other* – This also occurs in a minority of patients but is not too uncommon and can be something that easily leads to a very disgruntled patient. In more protracted cases that do not respond to reprogramming (which is always the first line of solution, and the second, third, fourth, and fifth), one should consider moving the lead. An anteroposterior and lateral X-ray of the thoracic spine should be obtained in these cases. If the rib or abdominal stimulation is always on one side, the lead may need to be moved closer to midline. If the unwanted stimulation is on both sides, the lead might work better if it is moved inferiorly a level or two. This is especially true of the lead is at T7 into 8. Nerve roots may come off the cord more medially at higher levels, and one may be able to avoid stimulating them at a slightly more inferior position.

Chapter 6

Paddle lead placement in the cervical spine region

Jeffrey E. Arle[1,2]

[1]*Neurosurgery, Harvard Medical School, Boston, MA, United States;* [2]*Neurosurgery, Beth Israel Deaconess Medical Center, Boston, MA, United States*

Scenario

A 52-year-old man is referred for placement of a cervical spine paddle lead after a successful trial by a pain physician in the community. He has had two anterior cervical discectomies with fusion in the past, at two separate levels (C45 and C56) at two separate times (3 years apart — first C56 and then C45). Although there had been improvement in his C6 distribution arm pain following the first surgery, he continued to have uncomfortable mechanical neck pain and then developed extension into his opposite shoulder over the subsequent year, eventually leading to the second surgery when the adjacent level C45 disk protrusion was noted. However, after this surgery, he had unrelenting neck pain despite what appeared to be a solid fusion construct and radicular-type pain on both sides in various distributions consistent with C5 and C6 and even C7 at times. After reviewing his films and noting no significant disk protrusion at previous levels or new levels, stable flexion-extension views showing a solid fusion, lack of notable benefit from gabapentin at 1800 mg/day and inability to tolerate pregabalin, and lack of success with epidural steroids and facet blocks and that he reported benefit from the trial stimulation of over 65%, a decision was made to proceed with paddle lead and implantable pulse generator (IPG) placement in the operating room on the elective schedule. Of note, he had also tried a round of chiropractic care as well as acupuncture treatments and had become accustomed to taking 5—10 mg of oxycodone every 6—8 hours, although he admitted this only improved his visual analog scale by perhaps one to two points from a typical 8/10 to 6/10 or 7/10 on most days.

The plan was to place the paddle lead, from a company primarily relying on traditional paresthesia-based stimulation, at levels spanning C3-4-5. The goal of the stimulation was suggested to be for his arm pains, but in the trial,

The Neuromodulation Casebook. https://doi.org/10.1016/B978-0-12-817002-1.00006-7
45

some benefit seemed to occur for the neck and trapezial axial pains as well. The patient was brought to the operating room and intubated. Using a total intravenous anesthetic technique approach for anesthesia, he was turned into the prone position using a chest and bilateral hip cushions, and his head was able to be positioned neutral to slightly flexed, resting on a horseshoe head holder attached to the Mayfield bed attachment. The back of his neck and base of the skull region was shaved adequately to allow for the midline incision and so the dressing afterward would adhere well. A midline incision was planned from approximately the C2 spinous process down to just superior to the C6 or C7 spinous process. An IPG site was planned in the flank region, marking it in the preoperative holding area with the patient in the upright sitting position — posteriorly located, to avoid having to perform two procedures in different positions (if the IPG was instead to be placed in the anterior subclavicular area) but not as far inferior as the upper buttock to minimize long wire pulling and extension wires. Both the subclavicular and upper buttock areas are viable alternatives, however. Also, this patient had enough excess tissue in his flank region to allow for this without likely discomfort once healed.

Variation: *Pain was primarily in neck and trapezial area* – Axial pain symptoms are rarely managed as well as extremity pain symptoms, and without doubt, neuropathic better than nociceptive. Axial pain is often more difficult to ascribe to the neuropathic category. That said, patients can have benefit from spinal cord stimulation (SCS) in treating axial symptoms, and it is worth a trial. The trial tends not to be a perfect discriminator in the long term but does provide justification to proceed with permanent placement if successful. In this case, if the patient were to have no extremity symptoms and pain was localized entirely to axial neck and maybe some trapezial elements and a trial was successful (i.e., >50% improvement for several days, perhaps better sleep or function, and worsening again once removed), then a paddle placement is likely the best option. In terms of location, such pain is more likely relatable to higher segmental levels than C4-7, which is represented in this case. In such a scenario, it would be more successful to place the lead from C1-3 if possible. Some have advocated placing a retrograde lead down from above C1. This is also reasonable but provides no significant advantage in my opinion to placing the lead from a lower level and sliding it superiorly; it can migrate or move from either position and needs to be anchored well in either place. In some cases, if a lower level cannot be accessed, the superior levels above C3 may yield lower segment pain benefit, even including low back or leg, but this is quite unreliable and rare in my experience. Even covering the hand is not reliable at all above C4.

Variation: *Pain is primarily in the hand* – As mentioned previously, pain predominantly in the hand, in failed back surgery syndrome (FBSS) or complex regional pain syndrome or other etiologies, is not reliably covered by paresthesia-based stimulators above the C4 level and ideally the lead needs to be at C5 or below. Placing it at C7-T1 may be too low and also unreliable. One might also consider peripheral nerve stimulation (PNS) over the distal median or ulnar nerves if SCS is unsuccessful.

Variation: *Pain is in the hand and neck* – If pain is in hand and neck, in other words, the most extremes – axial superior and distal extremity – then it is unlikely the patient will benefit with a single lead. In my view, the trial can be used in many cases to separate out the appropriate levels and may need to be considered as the permanent solution for therapy. If the patient can distinguish one as being more important than the other, that target should be treated first, but with the caveat that an inferior lead (e.g., C5-7) to treat hand will almost never treat neck pain, whereas a superior lead (e.g., C1-3) *might* treat hand.

Variation: *A nonparesthesia system is being used* – High-frequency (i.e., 10 kHz) and so-called Burst systems or subthreshold ~1 kHz systems might be planned from the outset, instead of a "traditional" stimulation system used in this case. All of them can achieve successful pain benefit. However, it should be appreciated that it is less clear where the lead needs to be placed. My own personal view currently is that the lead position should be planned as if it were going to be paresthesia-based and tested as such in the operating room (OR) as noted in the case here. Should the other waveform fail, at least the lead is positioned to try paresthesia-based therapy anyway.

Scenario (continued)

After an appropriate time-out, and with established motor-evoked potential baselines, the cervical incision was made, and dissection was carried down along both sides of the laminar margin using a small periosteal elevator and monopolar cautery. Fluoroscopy may be used but in the anteroposterior direction (to appreciate mediolateral aspects of the lead position), with the Mayfield in use, there is often difficulty getting reasonable visualization. As intraoperative stimulation and electromyography (EMG) is used and the spinous process of C2 is easily palpated or seen directly, the need for fluoroscopy in the cervical spine is quite minimal. Once the C2 spinous process is appreciated and uncovered on its inferior margin and the next several levels cleared to the bone margin as well, an initial partial laminectomy is made approximately at the C5 inferior margin, removing an angle of spinous process, lamina, and ligamentum flavum with rongeurs, the high-speed drill, and Kerrisons, thereby exposing the dura in the midline and for several millimeters out to either side to provide generally midline access for the lead. An motor evoked potential (MEP) is checked and compared to baselines.

<u>Variation</u>: *Cord signal is noted following earlier surgeries* – Most likely, if there is evidence of damage within the spinal cord, either attributed to previous compression and edema with the disk protrusions, trauma, or damage possibly related to performing the previous surgeries, the pain is only potentially related. It is entirely possible that there is cord damage and no chronic neuropathic pain comes of it. That said, it is extremely difficult if not impossible to disentangle the sources of the pain — whether from within the nerve root, dorsal ganglion, spinal cord, or nerve endings within the disk margins, facet joints, or interrelated muscle spasm and fascial inflammation. The nature of the pain from within the spinal cord tends to be more dysesthetic — burning, tingling, electrical, and associated with numbness that is disturbing or any combination thereof. In any case, the ability to provide relief of even this cord damage–related pain (by definition neuropathic) using spinal cord stimulation is high and should be tried just as much as other etiologies are considered. Placement of the electrode should be about one to two levels superior to the cord signal region. It is important not to place a paddle lead in an area which is already narrowed or inadequately decompressed, as may be the case in some situations where there is cord signal. In such cases, it would be reasonable to perform more of a full laminectomy over the area rather than trying to slide the paddle under the ligamentum flavum for the entire distance. Also, consider other variations as described in one of the answers below for laminectomy variants.

<u>Variation</u>: *The patient had prior posterior decompression(s)* – It is critical to review the previous bony removal with preoperative imaging prior to taking the patient to the OR for a paddle lead placement. There are two reasons for this. First, if there is prior removal of bone and ligament from the area where the lead is planned, it is frankly unlikely there will be much chance of placing the lead there without a significant risk for a dural tear or cord injury. Typically, if there has been extensive removal (e.g., laminectomies from C3-6), the scar extending down to and merging with the dura becomes extremely difficult to dissect without damage to the dura or worse. Most often in these situations, a percutaneous lead placement for trial cannot be done either and the process of trying SCS at all ends there, as no trial, percutaneous or paddle trial, can be performed. However, often the area of interest for placing the paddle is just *near* prior surgery but not

within it. Sometimes retrograde access can facilitate placement of a lead under these circumstances. Sometimes a small removal of scar can be performed safely and the lead can be adequately positioned. These are all worth considering, but the situation needs to be appreciated at the earliest point before any plan is formulated.

Variation: *The patient had previous posterior fixation but not full laminectomies* – As is also true with fixation and lateral mass fusion in other areas of the spine, if the lamina and/or spinous process is still in place, the tissue planes extending down to and including the dura are essentially normal, allowing for normal dissection and placement of a lead. In some cases, there is some scar or adhesions within the epidural space from prior lateral surgery, but in my experience, it is rarely an issue. Even in cases where the spinous processes had been removed and bone graft had been placed over the intervening laminae to enhance the overall fusion, if the original lamina had not been removed and the ligamentum flavum had remained intact, then drilling through a central region of the now-fused central area is very possible and typically allows for normal placement of the lead.

Scenario (continued)

The lead itself is then placed through the laminectomy opening which typically spans approximately 1.5 cm square and has been made hemostatic with bone wax, the bipolar and gelfoam soaked in thrombin as needed. The lead is placed from inferior to superior. The MEP is checked again. However, at this point in the procedure, placement of the lead can become difficult because the lead will not readily pass through the narrowed levels aligned with the disk margins superiorly. This is a common hindrance in the cervical spine despite the shorter distance between levels. Although narrower paddle leads (e.g., 2×8 configurations) require less width, overall length and structural rigidity are more important aspects of cervical lead design. If the paddle is longer, anatomical constraints do not let it pass up several levels as easily, and if the paddle is more pliable, it can curl or limit its ability to maintain its position within the epidural space as well while it is gently guided. Paddle leads have the advantage of being virtually immovable once they have fibrosed into place, and they flatten the dura toward the spinal cord more than percutaneous leads, minimizing (although not always eliminating) positional effects. However, it is important to understand the problems with ease of paddle lead placement.

In this case, the lead passed one level but was hung up for several attempts at the second level. Further removal and reattempting to pass the lead eventually led to it going all the way in. Testing with EMG stimulation using the four corners of the lead as bipoles showed that the lead was angled toward the right at the top and just left of midline at the base where it entered the epidural space. The lead was removed again, and attempts were made to reposition it more aligned straight and at midline as the patient had symptoms requiring bilateral coverage. MEP is checked. After two more attempts, the top of the lead was right of midline but clearly had contacts which could cover the left side, and the base of the lead was more or less at midline symmetrically. EMG thresholds at the base showed the left side came in first but the right side soon followed. Another MEP is checked here.

Variation: *The lead will not pass more than one level* – This is a critically important topic and applies as well to other areas in the spine where paddle leads may be placed (e.g., thoracic), but it is particularly common in the cervical spine. Because using too much force to pass the lead across the narrowed (and not directly visualized) region of the disk margin can result in cord damage and still may not allow the lead to pass, it is incumbent on the surgeon to decide when to abandon making only a single access to the epidural space and move on to several of these alternative strategies. The surgeon should become comfortable with these other options so that they are not overly driven to force the lead in with their only opening or abandon the surgery prematurely. The first choice is to make an additional opening similar (although not necessarily as long, but usually just as wide) as the initial opening one level superior to the original opening. This will enable manipulating the lead as it comes up from the level below, using perhaps a Penfield 4 or nerve hook to guide the paddle away from the lateral recess and allow it to move centrally into the sublaminar region of the next level. Even a second additional opening may be needed to help guide the lead and align it further. I have found it is not uncommon that cervical paddle leads require more than one laminectomy — particularly with some of the longest paddles on the market. This option is most specifically needed when the lead will not pass easily but also will not pass and remain midline.

In cases that are refractory to *any* passage, one can of course remove the lamina and ligament entirely, suturing the lead to the dura. However, it is helpful to bring the lead closer to the cord to minimize positional changes, but the help secure the lead better, thereby maximizing the best advantages of a paddle to begin with and this requires that some bone remain in place. To accomplish this kind of solution, one can remove the spinous process with approximately a centimeter-width opening down to the dura, and by removing ligament below the bone, as well as some ligament that remains under the lamina laterally to either side (often with a combination of a Leksell rongeur, drill, and Kerrison rongeurs), the lead can be slid relatively easily from the original access opening below, under direct visualization, as the lateral laminae still serve to hold the lead in place. One may also use variations of this idea, by removing only a hemilamina for example.

Scenario (continued)

After copious irrigation with bacitracin saline, the silicone anchors were placed over the lead wires and down to the region of the interspinous ligament. No reasonable amount of ligament was able to hold a suture of 2-0 silk adequately, and a penetrating towel-clip (an Edna) was used to make an opening through the spinous process at an appropriate level to pass the suture through the bone of the inferior process. The wires were secured there individually, and then the retractor was removed and the soft tissues made hemostatic with the bipolar cautery. The fascia was then closed with interrupted Vicryl allowing the lead wires to emerge near the inferior portion of the fascial closure. Another MEP is checked. The IPG pocket was created with a scalpel in a slight curvilinear fashion, and the pocket itself created partly with blunt finger dissection and the use of curved Mayo scissors and it was packed with gauze. The tunneler was then passed between the incisions, taking care in this case to pass medial to the scapular border but lateral to the paraspinal muscle margin. The wires were brought through the tunnel, dried with gauze at their ends and connected to the IPG, secured with the small hex wrench, and then placed into the pocket after removal of the gauze and checking for hemostasis within the pocket. Typically, impedances are checked with the connections and the wires removed and replaced before final torqueing if necessary. The two incisions are then closed with interrupted 2-0 Vicryl and either Nylon or subcutaneous monocryl for skin, although other closers are perfectly acceptable. A small loop of the wires under the 2-0 Vicryl is typically placed as a further strain relief. Care must be taken also to place the loops of excess wire in the pocket deep to the IPG and to make sure the IPG itself is facing (lettering) out toward skin. Final MEPs are checked again.

Dressings are placed, and the patient is turned carefully back into the supine position for awakening and extubation. Some care should be exercised in turning the patient so that the body is not twisted too much in the timing of the turn. Such movement, if extreme and in the right part of the body, could produce displacement of the lead itself in the epidural space. The patient is checked for movement in extremities and cared for in the typical way in the postanesthesia care unit. Often, the posterior cervical incisions are painful and patients need notable pain medication in the first few hours. Approximately 95% of these patients can be discharged from the postanesthetic care unit (PACU) the same day, however. Ketoralac is also often helpful in these patients with postoperative pain control. Most patients, as well, are able to be initially programmed in the PACU area by the company representative with several initial program settings to try when they are discharged. Also, instruction can be given about how to charge their IPG or communicate with it and change settings or turn the device on and off.

Variation: *Etiology was a brachial plexus avulsion injury or a thoracic spinal cord trauma with postparaplegic pain* – Both of these cases may benefit tremendously from SCS therapy, but as described previously in other variation sections, one must be aware of both the level needed for targeting (typically one to two levels above the area of injury) and the ability to access that region and place a paddle if there has been prior surgery within it or scar from trauma. Often, these types of etiology do not allow for a percutaneous trial, and so a paddle trial must be considered. This is described in a separate case study chapter in this text.

Variation: *Progressive myelopathy symptoms develop a year after lead placement* – This is a rare complication but can occur with any electrode placement in the epidural space. If it occurs, however, it is more often with paddle leads than with percutaneous leads, primarily because there is more surface area for fibrosis to form from and build up. Also, as a paddle lead is pushed somewhat more toward the dura and cord overall, as scarring increases in these rare cases over time, it is not impossible for significant cord compression to develop, thereby leading to myelopathy. Any patient who begins to develop symptoms consistent with myelopathy after placement of a paddle lead needs to be taken seriously and ideally imaged within a short time of verifying the symptom development. Often, paddle leads and the IPG have become MRI compatible. However, even an MRI may not show the region right under the electrodes in enough detail because of localized metal artifact. A myelogram and a post-myelo CT are still very helpful and should be considered if there is no MRI compatibility or if the MRI is simply not able to answer the question as to whether the lead and excess scar is compressing the spinal cord. Often, even when imaged, there is no significant compression and symptoms are from other etiologies or not able to be determined. If cord compression does exist and the SCS system is helpful for the patient, there are several options, but the best option may be simply to fully decompress the electrode by removing bone and ligament over its entire length. This laminectomy allows the lead to remain in correct alignment and location (rather than trying to move the lead to a new position) and allows it to move back away from the cord, relieving pressure and, presumably, the myelopathy.

Scenario (continued)

Reviewing the follow-up care given at the end of the case chapter for placement of a thoracic paddle lead for FBSS covers all the appropriate concerns that apply to the cervical paddle leads (see Chapter 5).

Chapter 7

Paddle lead trial in the case of a prior fusion

Jeffrey E. Arle[1,2]
[1]*Neurosurgery, Harvard Medical School, Boston, MA, United States;* [2]*Neurosurgery, Beth Israel Deaconess Medical Center, Boston, MA, United States*

Scenario

A 49-year-old woman is referred for placement of a spinal cord stimulator from a physiatrist who sees many patients with persisting back pain and failed back surgery syndrome in general. Although this physician does place trial leads in some patients, he typically is comfortable only with fairly straight-forward placements and does not do any permanent placements or implantable pulse generator (IPG) replacements. In this patient, there had been three previous surgeries. Initially, approximately 8 years before, she had a micro-discectomy, somewhat successfully at L5-S1 on the left. Unfortunately, she developed further problems within a year, back and leg pain with degenerative changes at L45 as well, and this led to a fusion from L4-S1. The patient did reasonably well for several years, although she began having more back pain and some bilateral leg symptoms 3 years ago when she was involved in a major motor vehicle accident, wherein she subsequently had a significant burst fracture and instability at T11, requiring fusion from T8 extending down to and including the L4-S1 fusion. Although this stabilized her spine and the construct appeared reasonable by all imaging since then, she remained compromised with worsening low back pain and bilateral leg pains, left worse than right.

The patient came to the visit with an magnetic resonance imaging (MRI) scan that was about 19 months old and several sets of thoracic and lumbar X-rays, clearly showing the extensive hardware construct. She had had multiple trigger-point injections, use of a TENS unit, and several series of physical therapy visits to strengthen her core. She was overweight with a body mass index of 30. She had been out of work since the car accident and was receiving disability payments. Her original job had been as a store manager for a large-chain department store. The burst fracture and subsequent injury appeared to

The Neuromodulation Casebook. https://doi.org/10.1016/B978-0-12-817002-1.00007-9

have left her with only modest neurological deficits, and she had been fortu-
nate in that regard. She had some initial bladder dysfunction, which had
recovered, and she felt burning and dysesthetic changes in her lower extremities which came and went over periods of minutes or hours at times. She
had no clear weakness focally on examination except related to effort to some
degree, giving 5 or 5-/5 at all muscle groups, although she did endorse having
some balance problems after the accident which had somewhat recovered after
physical therapy (PT).

She now felt that she could not enjoy any activities without significant
lower lumbar back pain which increased with activity from a baseline 7—8 of
10 to 9—10/10 and could only walk or stand for about 10 minutes at a time
because of pain developing more in her legs and her back as well. She did not
uniformly get relief with sitting and often had to go into her bedroom and lie
down for an hour or more to get any change back to baseline. Medications at
this juncture included gabapentin 2400 mg/day, ibuprofen 800—1600/day,
hydrocodone 20 mg/day, although it had not changed in over 2 years, and
methocarbamol 750 mg bid.

Variation: *Only had bladder dysfunction and numbness in lower
extremities* – Spinal cord stimulation (SCS) is not going to be of significant benefit for these neurological sequelae. It would be best to tell
the patient this in a direct way. Sometimes patients who have been
referred for SCS from their long-term pain management physicians
have a harder time believing this response and may doubt whether or
not the surgeon has understood their circumstances.

Variation: *Only symptoms of weakness in lower extremities and some
higher thoracic back pain* – The higher level back pain is much more
likely related to the soft tissues that had been disrupted with the accident and weakness from cord injury in the accident as well. Neither of
these symptoms is likely to benefit significantly from SCS.

Scenario (continued)

It was clear from the nature of the injury at T11 that a decompression had been
performed there and therefore the ability to pass a percutaneous trial lead from
below (typically entry to the epidural space is between L2 and L4) up to
approximately the level of T8, where a permanent lead would be best located,
would be likely impossible and not worth even trying because of the scar
adherent to the dura at the T11 region. This situation, however, does not rule
out the use of SCS in this patient. What could not be gleaned as well from the
prior MRI was whether laminae were left intact above T10. It appeared they
were, but the MRI was not the best study to examine this question, nor were
the X-rays ideal. They showed loss of spinous processes up through T8 and
density consistent with fusion mass over the midline, but there was no way to

confirm if the laminae at T8-10 had been removed. If the lead was to span from the bottom of T7 past the T8-9 disk space (given the length of most paddle leads), then it is still possible a paddle lead could be placed.

> **Variation**: *The decompression had been done from T7 through T10 –* This amount of decompressive scar attached down to the dura would likely prevent any possibility of placing a paddle lead. However, it could be that the patient has predominantly foot or more distal lower extremity pain. This could allow a lead placed just below the T10 margin for benefit.

> **Variation**: *The prior fusion was from T4-8 –* This would surely allow a lead to be placed below T8 and could provide substantial pain relief, covering perhaps the T9 level and especially the requirement currently for HF10 therapy to have the lead at the T9-10 disk space level.

Scenario (continued)

A discussion of the risks and benefits of SCS therapy took place as well as a discussion of how a paddle trial might work. Critically, however, a computed tomography (CT) scan was needed to determine what bone had been left in place during the prior surgery. A thoracic CT scan was ordered, and the patient had this done within a few days of that visit.

The following week when the patient returned to the clinic, the CT was reviewed and found that, importantly, the laminae had been left more or less intact above the area of decompression which spanned from the bottom of T10 to the top of T12. However, the spinous processes had been removed and bone graft extended over the midline regions of the laminae in the large fusion mass that had developed.

It was determined that a paddle could be placed in a reasonable position at approximately T8. It could be accessed either from below at approximately the T9-10 region (i.e., by going through the fusion) or from above the prior fusion and passed retrograde from an opening around T7. Although accessing the epidural space above the fusion would have been readily appreciated without a CT scan, the ability to pass the lead inferiorly might still have been quite difficult or impossible without knowing the bony anatomy left in place previously.

> **Variation**: *Cross-links in the hardware were located right were access was planned –* This situation is occasionally encountered in these relatively rare cases. One should have the confidence that working in and around and under the cross-link can be done without too much difficulty. More bone work might be necessary. However, a shortcut might be to consider that the region is already fused and the cross-link likely has no structural bearing anymore — as such, it can be damaged

by the drill or simply removed entirely without worry. Sometimes it is removable using just a Leksell rongeur or Adson rongeur, or if need be, a carbide drill bit and the high-speed drill can easily remove it.

Scenario (continued)

Once I had determined both the ability and the appropriateness for SCS therapy, a discussion was again had with the patient to explain how a paddle trial is performed. Two surgeries approximately a week apart are scheduled. I explained to her that the first surgery is to place the paddle lead under typical circumstances (anesthesia, OR imo, electromyography [EMG]-stimulation testing), as if it were a permanent lead placement. The lead is secured as if it was permanent, and the wires are brought through the fascia layer, closed in the usual manner with interrupted 0-Vicryl. However, outside of the fascia, trial extension wires are connected to the ends of the lead wires and then tunneled out through the skin laterally away from the incision. These are connected to a stimulator which allows the trial to then proceed in a typical fashion. The second surgery is either for placement of the IPG or removal of the lead if unsuccessful. The patient agreed, and the surgeries were scheduled.

Preoperative assessment and preparation take place as previously described. Importantly, the scheduling of the cases necessitates that the patient be prepared to be discharged after the first surgery and go through the typical assessment of the trial period and then a second surgery a week later. In some ways, the number of procedures is similar to having a percutaneous trial lead placement followed shortly later by permanent lead and IPG placement, except that it is more of an invasive surgery per se.

She is seen in the preoperative holding area on the day of surgery, and the previous surgical scar is appreciated extending high up into the mid-thoracic area. There is evidence of her having had multiple lower back midline incisions. I decide to access the area of interest from within the fusion and try to place the electrode inferior-to-superior and so I place a mark on her back near the upper region of the incision. She is brought into the operating room, intubated under total intravenous anesthetic technique (TIVA) anesthetic, turned into the prone position with appropriate chest and hip bolsters, and prepped for this midline thoracic reexploration. Prophylactic antibiotics are given once she is asleep (typically 2 gms of intravenous cefazolin unless there is an allergic concern). Instead of including an upper buttock IPG site in the field, a wider region at the level of the incision is prepped to allow for the trial wires to be tunneled out. After an appropriate timeout, the incision is made and dissection is taken down directly to the bony landmarks, clearing them from scar and other tissue primarily using the monopolar cautery and a cerebellar for self-retaining retraction.

Eventually the titanium rods and two levels of the upper pedicle screw heads are able to be identified. Bone graft fusion extends across the midline, and part of the rods are enveloped in some of this bone. Using lateral fluoroscopy, and the construct landmarks, the appropriate entry level is determined, and the high-speed drill is used to make an opening in the bone graft between the rods in the midline region. Careful drilling proceeds until the ligamentum flavum is encountered and then hemostasis is obtained using bone wax and the bipolar or gelfoam in thrombin as needed. The opening is widened to a typical size, and removal of the ligament proceeds using primarily Kerrison rongeurs. Care is taken not to disrupt too much of the fusion graft, thereby making the construct unstable or potentially creating a pseudoarthrosis. Motor evoked potentials (MEPs) are checked here and periodically throughout the surgery. Once adequate access to the epidural space has been made, the paddle lead is gently placed within and advanced as able superiorly. Most of the time, in cases like this, the lead advances or encounters hindrances as in a normal surgery. In this case, the lead was able to be placed and aligned appropriately using fluoroscopy and EMG-stimulation testing after the second placement attempt. No additional laminectomy levels or retrograde access was required, although those would have been reasonable strategies if necessary.

Once the lead was positioned, it was anchored in the usual way although with interspinous ligament to work with. In this case, a small passing hole was able to be made (using the drill and penetrating towel clips) within the fusion graft in the midline whereupon the wires and anchors were able to be sutured down. Copious irrigation with bacitracin-infused saline was performed, and the retractor was removed. Hemostasis was obtained in the soft tissues, and then the fascia was closed in the usual way with 0-Vicryl. Care was then taken to note how much room there was for the coils of excess wire and the connectors for the trial extensions in the suprafascial space. Using the monopolar cautery, the subdermal fat tissue was freed up more to either side to make a slightly larger amount of room to contain these wires.

The extension wires were connected to the ends of the lead wires, and then the tunneler tunneled approximately 4 inches lateral to the incision where the tip was helped through the skin with a #11 scalpel blade. The trial extension wires were brought out to this site and then the wound was closed with interrupted Vicryl and skin sutures. Dressings also covered the extension wires almost to their contact ends which were left open to air. An additional gauze was left just under the wires where they exited from the skin, as this tends to drain more and longer.

She was programmed in the postanesthetic care unit (PACU) and sent home with pain medication and instructions on how to use the controller and how to contact the company at any time as well as my office. Over the next week, I received updates from the company representative, and then before the weekend, from the patient herself. She thought she had notable benefit in her legs but was n't sure the coverage extended high enough into her lower back. Another programming adjustment was made by the representative, and she said she had good relief in her low back and legs, up to 60% from her baseline, and she felt like she could walk and stand for significantly longer periods of time. She wanted to have the IPG implanted.

When I next saw the patient, the following week in the holding area again, her midline incision healed well and the drainage from the wire exit site had subsided and stopped about 4 days prior. Infection is always a potential complication in these staged surgeries, and with implantation of a device, it is important to be honest with oneself as to whether there is evidence of frank infection or not when the patient returns.

> <u>Variation</u>: *Pus is draining from the exit site and lower centimeter of midline incision when the patient returns* – Although quite rare, overt infection such as this prior to placing the IPG needs to be treated aggressively by removing the lead and all wires and anchors, particularly to minimize the risk of osteomyelitis. Antibiotics typically are used for a minimum of 2 weeks or longer and can be refined once culture results from the surgery returned. Salvaging the lead with infection at this juncture is not recommended – the trial information has been obtained and one can always return and place a new lead in the same location, or nearly so, and the likelihood of avoiding subsequent infection is low. That said, without overt pus, but just typical granulation tissue and serosanguinous fluids in the prior operative area, good debridement and irrigation is likely sufficient to leave the lead in place and add the IPG.

Scenario (continued)

Again, the patient is brought to the operating room, intubated although with any form of general anesthesia this time – there is no need for TIVA. Again, prophylactic antibiotics are given within 30′ of the incisions. We plan an upper buttock incision for the IPG on the same side as where the wires for the trial exit. This may seem to be counterintuitive, but as long as the exit site is prepped out of the surgical field and the leads are tunneled inferiorly and not near the site where the wires had exited, the infection risk is no greater than going to the opposite side. As such, the IPG is determined based on the typical concerns: which side does she typically sleep on? If male, does he wear a wallet? Does she have pains or prior surgeries on one side or the other? The median incision has sutures removed and is prepped as in the exit site and the entire region as in a typical permanent implant case where the IPG site is included. The drapes, however, marginalize the exiting trial wires out of the field. The circulating nurse apprised these wires and will pull them from the body from under the drapes once they are cut from within the midline operative area. Arms are placed bent and up toward the head.

After an appropriate timeout, the incision is made in the previous thoracic midline incision, taking care to remove any previously used 2-0 Vicryl and not damaging any of the wires that are immediately present upon opening this suprafascial area. Serosanguinous fluid is readily apparent and drains exposing early fibrous scar adhesions forming in and around many of the wires. All of this is aggressively suctioned and soft-tissue margins are carefully debrided to remove any potential early infectious elements as the wires are allowed to rise up in a twisted collection from the opening. Taking care not to pull the exiting trial wires into the wound from their externalized location, the appropriate wires are identified and disconnections made by using the hex wrench screwdriver. The connectors, along with a centimeter or so of wire, are then cut using a Mayo scissor, and the circulating nurse is instructed to pull the wires out from under the drapes. The fascia is left closed and intact. The IPG site is created with either blunt finger dissection or using curved Mayo scissors, just large enough for the IPG. It is packed with gauze, and the tunneler is then passed between the two incisions, the leads pulled through the IPG site and connected to the IPG after drying them off, and the IPG placed into the pocket after removing the gauze, making sure there is hemostasis within the pocket. Often, the impedances are checked to make sure the connection is sound. Both incisions are then closed using 2-0 Vicryl and 3-0 running unlocked Nylon to close the skin. Dressings are placed, and she is turned back into the supine position, awakened, extubated, and taken to the recovery room for further care and start-up of the stimulator. Most likely, she will be able to go home the same day after a couple of hours in the recovery room.

If the trial turns out unsuccessful, again the patient is brought to the operating room, intubated although with any form of general anesthesia this time — there is no need for TIVA. Again, prophylactic antibiotics are given within 30′ of the incisions. The drapes encompass only the midline thoracic incision. The exit site is prepped but draped out of the field, and again the circulating nurse is made aware that they will need to pull the wires out from under the drapes once they are cut internally. The median incision has sutures removed and is prepped. Arms are placed bent and up toward the head.

After an appropriate timeout, the incision is made in the previous thoracic midline incision, taking care to remove any previously used 2-0 Vicryl and not damaging any of the wires that are immediately present upon opening this suprafascial area. Serosanguinous fluid is readily apparent and drains exposing early fibrous scar adhesions forming in and around many of the wires. All of this is aggressively suctioned, and soft-tissue margins are carefully debrided to remove any potential early infectious elements as the wires are allowed to rise up in a twisted collection from the opening. Taking care not to pull the exiting trial wires into the wound from their externalized location, the appropriate wires are identified and disconnections made by using the hex wrench screwdriver. The connectors, along with a centimeter or so of wire, are then cut using a Mayo scissor, and the circulating nurse is instructed to pull the wires out from under the drapes.

At this point, the fascia is opened and the previously placed Vicryl sutures are removed. The anchor sutures are cut using a #11 blade, and the lead is pulled gently out of the epidural space. It has not scarred in long enough to be held in place to any significant degree. The lead, anchors, and wires are all removed completely. The deep and more superficial areas are copiously irrigated with bacitracin saline, and then the wound is closed in the typical multilayered fashion. The patient then has a dressing applied, is turned back into the supine position, awakened, extubated, and taken to the PACU for further care.

She did well overall with the surgery and healing. I removed her sutures on day 14, and there were no concerns with the healing of the two incisions or the exit site, except for a slight tenderness at the IPG site. The company representative was there at the visit to spend some time on reprogramming and go over the use of the controller and recharger again. Because the trial lead was also the permanent lead, she had fewer overall complaints early on that coverage was not the same as the trial. Nonetheless, I reassured her that it was still early and with time as the scar settled internally and some further reprogramming it was likely to be better yet. I also reiterated that she should be proactive in meeting and having further reprogramming as needed in the future.

Chapter 8

Complications involving spinal cord stimulation and implantable pulse generators

Jeffrey E. Arle[1,2]

[1]*Neurosurgery, Harvard Medical School, Boston, MA, United States;* [2]*Neurosurgery, Beth Israel Deaconess Medical Center, Boston, MA, United States*

Scenario

A 42-year-old man came into the clinic as a referral to reposition or at least evaluate his spinal cord stimulator system. He had a long history of several revisions which he described at some length after a series of probing questions trying to unravel the precise history of what was done, when, and why. Some of these records were sent from other providers, and although available, they did not necessarily clarify all the details of what had been done previously. Initially, he said "a single wire" was placed at the time of the trial, or just afterward, which he could not recall now. This never helped him at all for his back pain, and he remembers he felt something like a TENS unit in his leg but it was n't in the right area of his pain. That doctor, he said, told him that the lead had moved and "pulled out" or something and that he replaced it with two wires he thinks, but it required two different procedures somehow, and he was n't sure what was done during each of them. This revision helped his pain, eventually, for a couple years, but then one wire broke, he thinks he was told, and he did n't have any stimulation really covering the area where his pain has remained, despite several times seeing the company representative and having reprogramming done. More recently, perhaps in the last 6 months, his pain physician retired and his primary care physician (PCP) would not prescribe the same medications for his pain that he was previously obtaining from his pain doctor. He wanted to see if anything could be done with the stimulator or whether he should just have it removed.

> **Variation**: *The patient was referred instead for spinal cord stimulation "not working"* – Often patients come into clinic with spinal cord stimulation (SCS) problems and say "It's not working". This is a catch-all phrase that needs a degree of probing to determine the exact nature

The Neuromodulation Casebook. https://doi.org/10.1016/B978-0-12-817002-1.00008-0
61

of what is meant. This is critical for the simple reason that many patients have been informed, by other providers or friends and family, or have come to some conclusion on their own about what the problem is. So, they typically have some set of presumptions which needs to be worked through, and unless there is a fairly systematic thought process, the evaluation can become derailed quickly. In general, the algorithm involves determining a couple of important distinctions:

- Does the patient feel any stimulation at all? (assuming it is paresthesia-based)
- If not, it is important to determine whether they have *ever* felt stimulation, given the growing use of "nonparesthesia"-based systems, whether HF10 or subthreshold.
- If they feel stimulation, is it where they have pain?
- If it's not felt in the right areas, then reprogramming is needed.
- If it's felt in the right areas but does n't help their pain, little can be done except (if available) reprogramming with other frequencies and pulse widths, but this is typically less successful although not impossible.
- If it's felt in the right areas but the feeling is annoying or adds to their pain, then reprogramming can be helpful, but this may be a sign that the therapy now is not going to be helpful for the patient even if it had been in the past.
- If no stimulation is felt, does the device actually turn "on"? This type of evaluation can often be accomplished relatively quickly by the company representative at the clinic visit. They can also then determine expected battery life remaining, whether the device seems to have communication problems, and whether recharging (if it's rechargeable) is working correctly. They can also check impedances within the system and determine usually if there is a lead break somewhere or a complete disconnection.
- Sometimes, X-rays are helpful if a possible break or intermittent break is suspected, although it can be difficult to be sure if there is a break in a wire by X-ray alone, and often the end result of the evaluation is that the system has to be surgically explored and evaluated intraoperatively by directly testing the lead(s) alone, disconnected from the implantable pulse generator (IPG), no matter what is thought to be seen on an X-ray. In most of these cases, but not all, the lead and IPG need to be replaced — but it is worth verifying each of them before assuming that is the case.

Variation: *The patient is referred instead for a "new pain"* – If the patient begins by explaining that they think their pain is different than it had been or they have the same pains they always had but with additional new pain in a different location, it is important to distinguish whether the new pain is caused by the electrical aspects of the SCS itself or a new underlying anatomical development or potentially a mechanical change related to the stimulator itself (i.e., the lead, scar, or the IPG). A simple way to do this (which the patient may have tried but more often has not) is to turn the system "off" for an extended period of time and see whether the pain still occurs. This may seem obvious, but there is a certain amount of "washout" period necessary and the patient may have tried it for too short a time (usually several days are need to be sure) or the patient has been reluctant to turn the device off because it helps them with their underlying pain too much and they do not want to feel that pain back again. Many times, the patient has not actually thought of turning the device "off" at all but has also frequently come to the potentially erroneous conclusion that the device is the cause of their new pain. In any case, it is important to try to have this assessment made as a first step most of the time. Other components of the algorithm are as follows:

- If they still feel the same pain when the device is turned "off" and it is a pain possibly related to the spine somehow, then imaging may be needed to determine its cause (e.g., it seems to be a new radiculopathy). If the device is magnetic resonance imaging (MRI) compatible, an MRI can be obtained; if not, a CT myelogram can be obtained.
- If the new pain is more relatable to areas near the IPG itself, or the thoracic or cervical incisions, then it depends on how severe this pain is and the nature of it. Pain caused or exacerbated by palpation near these regions is likely due to pressure from the device components under the nerve endings within the dermis — this often occurs as scar matures and contracts and after notable weight loss. Lidoderm patches can be tried initially and are often helpful, but in many cases, they are not adequate, and revising the site in the operating room (OR), in some cases moving the IPG (for example) to an entirely different location, needs to be considered. Such circumstances are not uncommon, although certainly not in the majority of cases.

Scenario (continued)

A set of anteroposterior and lateral thoracic and lumbar X-rays were obtained which determined there were two percutaneous leads in place, one spanning from T10 to T11 and the other partially out of the epidural space in the L3 region with the other. There was nothing obvious that suggested a wire was broken in either system. He was also thinking that his back pain had gotten worse over the prior year or so. A communication with the company representative revealed what system he had in and that his battery required recharging more often than it should, which they surmised was also a part of the problem because he did n't recharge often enough and the battery would run low, providing inadequate levels of stimulation at times. They were also able to reveal some of the prior parameter changes that had been tried at times and some clarifying aspects of the history in terms of which leads were used when and what revisions had been done.

Variation: *Both leads appear to be intact and reasonably well positioned* – Electrical evaluation of the system needs to be performed, initially with an impedance check and then in the OR by opening the IPG site initially and then testing the lead itself directly. The IPG may need to be replaced, but the lead left intact, the lead alone may need replacement, or both may need to be replaced. In the case of two percutaneous leads, one or both may be able to be pulled down and out from the IPG site only, but a new lead (e.g., a paddle lead) will need to be placed anyway if the system had been helpful, and this requires a new incision, and the details of this placement was worked out beforehand with the patient (who consented for the procedure and was prepped in the surgical field ideally).

Variation: *The electrode is actually a paddle lead that is at T10-11* – The assessment of the lead intraoperatively may still need to occur and can be accessed in the same manner by opening the IPG site first and disconnecting the lead wires from the IPG and connecting them to the testing cables. However, additionally, it is important to understand whether (1) the therapy has helped the patient adequately in the past at all, and (2) if not, whether the lead itself was not ideally positioned and needs to be revised; in this case perhaps superiorly because it was never getting adequate low back coverage (or perhaps inferiorly to T11 or T12 if foot coverage was needed and never obtained).

Variation: *The electrode is a paddle electrode with some contacts showing high impedances* – This situation does occur occasionally, and most of the time, it is worth reprogramming the lead to work around the presumably broken contacts (or there is a tiny amount of fluid in the connection at the IPG which disables some contacts and not others). Often, the contacts have been dysfunctional for a while, but the programming for the patient had not required using those contacts anyway. If too many contacts are not working, in all likelihood, the lead will need to be replaced, and then it is important to assess with the patient whether it has been ideally positioned to begin with.

Scenario (continued)

It was clear from the X-rays that one lead obviously had migrated down and almost out of the epidural space altogether, likely leading to the loss of coverage, or at least most of the coverage. It could be that the other lead had migrated as well, but earlier placement films were not available to compare. Although it was unlikely that the single percutaneous lead still at T10-11 would provide all the relief he needed, he had n'ot been seen and reprogrammed since the loss of coverage. Seeing what this single lead would yield for coverage if optimized would be the best initial beginning to the evaluation — noninvasive, assesses the status of the current system and helps determine the efficacy of SCS as a therapy overall for him. This would be a first step, I told him, but subsequently it was entirely possible he would need both leads removed and a new paddle lead placed higher at approximately T8, as this would have the best chance of covering *both* his back and both legs and maintaining it for the long term. A new, updated IPG might be necessary as well.

A programming session was set up within a couple of weeks, and when the patient came in, he also noted that whenever he sat down, the IPG site pushed in and bothered them. It sometimes also bothered him when he walked, but mostly when he sat. He said he had lost some weight over the past 2 years, and it seemed to make the IPG stick out more and touching it was sore. One of the important points of the history was that the SCS therapy *had* at some point been beneficial. This was a critical aspect of evaluating any patient with dysfunctional stimulation overall. Asking "has it *ever* worked to help your pain?" remains a baseline to determine whether removal, revision, reprogramming, and so forth make the most sense. For him, the stimulation had been helpful in the past with over 50% benefit for his legs but less than that for his low back, although he does think it had helped more in his low back early on.

> **Variation**: *Original trial never met 50% criteria — the patient was just told they "had to have this done"* – It is important to bring out details of the history in patients with SCS such as this. If one takes everything sent in as a referral at face value, it can lead to expensive and unnecessary dead ends of therapy and ultimately sullies the reputation of the therapy itself for unwarranted reasons. Some patients were implanted when they probably should not have been. Sometimes the type of pain and the location of pain are not likely to benefit from SCS, but the patient was implanted anyway, perhaps with an inadequate trial. They may have told the implanter what they needed to hear to move forward with the permanent implant placement — 50% or more benefit in the trial — otherwise they know they will not get the therapy. Many patients in pain are so desperate for benefit they will do almost anything to try. Many practitioners who are managing patients, perhaps prescribing all their pain medications, desperately want to transition the patient to a different therapy not involving addictive medications and potentially even out of their practice. They may tell the patient they must have the implant or they will stop prescribing medication. In either situation, or other variations on these themes, patients may end up with SCS therapy

that never was appropriately evaluated in the beginning. Figuring this out once the patient arrives now looking for a revision is clearly needed.

Variation: *The trial was fine, and the system worked for a while (1 + years), but then before a revision was needed, they said it was n't helping pain anymore although stimulation was in the right place –* This development is also a critical decision-making point that must be determined when assessing revisions and replacements. Sometimes patients will come in saying "it doesn't work anymore", and what they mean is that the system does generate appropriate current and is technically functioning otherwise well, but the stimulation no longer provides pain benefit even though it had been beneficial and is still felt in the correct areas. This so called "tolerance" to SCS has an unclear etiology and is an active area of current research. Patients often will say they "can feel the stimulation and their pain at the same time independently." Possibly, there is an eventual imbalance in the dorsal horn circuitry that led to inhibition of wide dynamic range neurons carrying pain, now preventing such inhibition or in some cases enhancing their activity. In general it is worth turning the system "off" for about a week and trying to turn it "on" again and reprogramming it with different frequencies and pulse widths. If this does not work, it is likely that SCS will not work further and the system can be simply turned "off" or removed.

Scenario (continued)

When the single lead was accessed, it only could provide stimulation to one leg and could not reach into the low back at all. In all likelihood, it had migrated as well, but we could not be sure. Some of the stimulation he did feel in his leg was helpful for pain there, and this continued to suggest that the therapy in general was likely to be still helpful for him if the system was positioned well. After some discussion confirming he wanted to try to continue to use his system if it would help, we decided to schedule him for removal of his current SCS leads and IPG and replacement with a new paddle lead and IPG at the T8 level. This surgery was completed without incident, and he achieved stable 40% improvement in his lower back pain and 75% improvement in his leg pain to date.

However, after the replacement of his system, several months later, he returned with complaints again about soreness at the IPG site. It was clear the tissue between dermis and IPG had reduced and scar contracted, making the IPG more prominent there. Lidoderm patches were tried and failed, so an operation was scheduled ultimately to place the IPG deeper, rather than move it to a new position. In the OR, the deep fascial margin of the pocket was opened along the superior margin, and a new pocket was created deep to this scar. The IPG was placed into this new pocket, below the scar, and the margin closed with Vicryl suture. The old pocket as such was then tacked back together with several 2-0 Vicryl as well to prevent the creation of a seroma, and then the incision was closed with interrupted Vicryl as well, followed by skin.

> **Variation**: *The replaced and revised IPG became painful and sore after several months* – Occasionally, a patient will have soreness develop in and around the IPG pocket. Despite treating it locally and then revising the pocket, placing the IPG deeper (as described previously), the area remains sore and annoying. Even if the stimulation provides pain relief, the IPG area may be aggravating enough that the patient still wants the device removed. It is also worth noting that even if the IPG is moved to a different site (e.g., from one buttock to another or to the abdomen, or from the flank to the subclavicular area), the new site may be just as painful, *and* the old site may remain painful. In general, this is not always avoidable somehow, and there is no other known option to repositioning the IPG except removal.

> **Variation**: *The lead could not be repositioned at T8 because of other surgeries or scar tissue* – As noted in previous sections, prior surgeries and the scar that extend over the dura may prevent placement of the electrode safely in a desired area. Other locations or types of therapy should be considered in such cases.

Chapter 9

Dorsal root ganglion stimulation

Timothy Ray Deer

The Center for Pain Relief, Inc., Charleston, WV, United States

Scenario

A 21-year-old woman presented in referral for treatment of a severe lower extremity pain. She had initially been treated after a horseback riding injury with a nondisplaced fracture of the fibula. Despite good bone healing, she had persistent pain of the limb and developed coldness to touch, skin discoloration, atrophy of the skin, shininess and excessive hair growth, and hyperhidrosis. She was diagnosed with complex regional pain syndrome (CRPS), type I. She did meet all criteria for this diagnosis and had been treated appropriately with physical therapy, lumbar sympathetic blocks, anticonvulsants, nonsteroidals, and low-dose opioids. She was a college student and did not do well with medications because of side effects and did not respond long term to injection or physical therapy. She had an electromyography/NCS which showed no abnormalities and had a three-phase bone scan that was inconclusive.

On evaluation, the patient had pain from her knee to her toes including the bottom of her foot. She had allodynia and obvious changes in her skin, hair, and nails. At this time, I confirmed the diagnosis of CRPS type I and discussed treatment options. We did extensive patient education and considered different stimulation options including burst, high frequency, and tonic. After this discussion, we went over the option of dorsal root ganglion (DRG) stimulation and the United States investigational device exemption (IDE) study that led to the approval. Her history and presentation was consistent with the patients in the study. Based on this discussion, she opted to move forward with a trial of DRG stimulation. Based on her pain distribution, we opted for a lead to be placed at L4 and L5.

Variation: *The patient has a disuse injury from lack of activity and weight bearing* – In this setting, the use of any form of stimulation may be overly aggressive. Options would include a tunneled catheter for a few weeks to increase tolerance to therapy, a tunneled stimulation trial,

The Neuromodulation Casebook. https://doi.org/10.1016/B978-0-12-817002-1.00009-2

or potentially a 30-day cognitive program. In some of these settings, the placement of a permanent DRG stimulation device will lead to a reduction in pain levels and a tolerance to exercise. If the patient responds in this fashion and can undergo extensive therapy and desensitization training, they can often return to a normal status. In this setting, the device can be removed. We generally recommend a 6-month therapy hiatus to assure there is no relapse, before the explant.

Variation: *The patient has a peripheral nerve injury as the primary issue* – In some settings, a patient will have an injury of a peripheral nerve such as the superficial peroneal, tibial, or sural nerve and subsequently have an allodynic limb that develops signs of atrophy from disuse. In this setting, the use of peripheral nerve blocks and oral medications often fails. In this patient population, there is also a very low success rate of topicals, bracings, and alternative methods. Patients in this group can be treated with peripheral nerve stimulation (PNS) of the primary pain generator. Difficulties can arise from difficulty wearing a radiofrequency device in the lower extremity, and also problems may arise in devices that must couple to a Bluetooth system. Because of the limitations in these current devices, we would still move forward with DRG stimulation in the majority of these patients with the diagnosis of primary nerve injury or causalgia. The targets would be L5 or S1 for the foot, L4 or L5 for the area from the knee to the top of the foot, and L3 or L4 for the knee and surrounding structures. An alternative could be spinal cord stimulation (SCS) particularly if the anatomy does not merit a PNS device or DRG system.

Variation: *The patient has more widespread symptoms consistent with a radicular component in addition to the peripheral nerve injury* – Just as the patient with primary nerve injury, someone presenting with more than one pain generator is more complex. In this setting, after trying the conservative therapies noted in this discussion, additional workup is needed. A magnetic resonance imaging scan of the spine should be obtained, and consideration of a curative surgical lesion should be discussed. Even in this setting, the CRPS must be considered and may lead to worsening with any surgery including spine surgery. In this case, we would consider SCS as a primary implant method, with DRG as a potential backup. This is important because recent studies have shown that DRG lead placement at T12 or L2 may give considerable relief of both discogenic back pain and postsurgical pain.

Scenario (continued)

The patient was in pristine health with the exception of the lower extremity CRPS. We discussed anesthesia options and opted to move forward with a general anesthetic with neuromonitoring to improve safety and potentially efficacy. The patient was seen in the preoperative area, and an IV was placed. Vancomycin was given preoperatively, and the patient placed under general endotracheal anesthesia and positioned in the prone position. Based on pre-operative examination, an L4 and L5 DRG lead placement was planned for the trial. The patient was widely prepped and draped. Timeout was obtained, and all equipment was found to be available. Even though general anesthesia was used, I placed 1% lidocaine with epinephrine in the path of the needles to lower the anesthetic requirement and reduce postoperative pain. A small stab wound was made on the contralateral side, and a 14 gauge needle was used to enter the epidural space at L4/5. The DRG sheath was directed to the target just below the pedicle at the L4 foramen. I was able to enter the target zone of the DRG within seconds. The technique involves using the guidewire in the sheath and then transferring the lead once in place. This approach was repeated at L5, entering the L5/S1 interspace and the L5 foramen. During both placements, the neuromonitoring was quiet with no activity on the electro-myography or the somatosensory evoked potentials (SSEP).

Once the leads were in place, a simple bipolar program was used to check for motor recruitment and to check impedances. The impedances were both below 1000 Ω, which is ideal for DRG. At this point, the needles, stylets, and sheaths had been removed and the leads were secured to the skin with 0-Ethibond and silastic anchors. Sterile dressings were placed, and the patient was taken to the recovery room. The total surgical time for the trial was less than 15 min.

The patient's pain level was minimal postoperatively. She was programmed with a paresthesia mapping using both leads to cover the entire pain distri-bution, then the program was reduced 30% from threshold. The patient was sent home with a subthreshold program and instructed she would be seen back in 5 days. The plan was to use a subthreshold program for 5 days and then either conclude the trial or extend it for five additional days with tonic if needed. On the return visit, the patient reported a pain reduction of 90%, with increased function, sleep, and a reduction of allodynia. The leads were removed in the office, and the needle sites were dressed.

Variation: *No response to subthreshold stimulation* – Studies have shown that many patients do as well or better with subthreshold stim-ulation, and this is a very low energy state. For these reasons, we usually start a DRG stimulation trial with that programming in a simple bipolar orientation. In cases where the patient has no response at day 5, we would consider a tonic phase trial. The patient is mapped in the recovery for coverage of the distribution of the pain pattern at the time of lead placement, so on day 5, if there is no response to the initial program, the patient should have appropriate coverage. In these situations, we normally continue the trial for a total of 10 days.

Variation: *The leads shifted during the trial phase at day 3 and stimulation was lost* – The leads are anchored to the skin with a suture or a piece of tape. The other method to hold the lead in place is the epidural S-loop which is placed at the time of sheath removal. In some settings, despite this strain relief, the lead can shift. This may be more common in the obese, those with difficult foramen, or in those with excessive movement. The placement of two or more leads may help dissipate the risk of movement, and in some patients, I have placed four leads for the trial and based the permanent placement on the pain response to programming. In the future, we are hopeful a paddle lead will be available, Food and Drug Administration approved, that will solve this issue in those prone to lead movement.

Variation: *The leads could not be placed at L4 and L5 because of procedural difficulty* – In some cases, the preoperative imaging will show us an anatomical variant that is not conducive to lead placement. Unfortunately, in some settings, the inability to successfully place the lead is not noted until you are in the operating room. This can be treated by varying the normal and planned technique. The options would include placing a lead at L3 and S1 to attempt to cover the pain distribution. The other option would be a hybrid trial. In this scenario, the lead is placed at the nearest acceptable lead, and then a spinal cord stimulation or dorsal column lead is also placed. During the trial phase, the patient can experience both methods and choose the most optimal therapy.

Scenario (continued)

The patient returned for permanent implant placement 1 month after the trial, the preexisting leads were removed and the permanent leads were successful placed at L4 and L5 without difficulty. The leads were secured with one 0-Ethibond suture at the level of the superficial fascia. A pocket was made just above the belt line based on patient preference and anatomical features. The procedure went well without complications, and the patient immediately showed a good response.

Variation: *The patient requests a monitored sedation anesthetic instead of a general anesthetic* – In 2017 I began a process of implanting all DRG leads under general anesthetic with neuromonitoring. While this is not a fail-safe method of avoiding injury, it is a potential ideal method for creating the safest clinical environment possible. If the patient prefers to be awake, then it is important to properly gauge the sedation level so as to assure feedback at critical times during the procedure. In this setting, we would do the permanent monitored sedation but would create a careful plan with the anesthesia care team.

Variation: *The patient has been taking fish oil and other supplements* – The American Society of Regional Anesthesia and Pain Medicine and the Neuromodulation Appropriation Consensus Conference have given great guidance on these issues. We would take additional history and consult the appropriate resources to adjust the procedure timing and medication based on the proper safety guidance.

Variation: *The patient has had a previous back surgery* – I do not recommend going through a surgical scar or area to place DRG stimulation leads. In this setting, the epidural space may be scarred or for the most part absent. This can create a risk of dural tear or nerve irritation. In many of these patients, the lead can be placed at S1 for the foot and at L2 or T12 for the axial back. In patients with CRPS I like the young lady in this case, I would consider a T11-12 SCS lead with a Burst DeRidder program.

Scenario (continued)

The patient had an excellent postoperative course. Her wounds were well healed, and postoperative pain was at a minimum. She presented to the office wearing a shoe and socks. This was the first time we had seen her in either, and her allodynia was very minimal. We limited her activity for 6 weeks and then asked her to slowly increase the physical therapy and daily exercise. Within a few months, she was back to her normal activities including horseback riding, hiking, and running. She continued to do well at the 2-year follow up and continues with a subthreshold program with very limited energy requirements. She is currently in graduate school and is on no medications with no limitations. When she turns the system off for more than a few hours, her pain escalates and she sees changes in her skin color and temperature.

Variation: *Lead migration* – During the increased activity, the risk of DRG lead movement is a real possibility. The proper placement of the S-loop for slack in the epidural space is felt to be a critical part of reducing this risk. If there was lead migration, we would plan a percutaneous revision. As there is no paddle lead option at this time, the percutaneous route is the only option to recover stimulation. In the event we could not recapture the same levels, we would consider L3 and S1. These would not be ideal, but in many cases, these would potentially capture enough fibers to create a good outcome.

Variation: *Lead fracture* – In this setting, the patient has extreme activity that can lead to a fracture of the lead. I have seen this in a similar case where the patient fell out of a truck and fractured the lead. This was verified by X-ray.

Variation: *All symptoms resolve in the lower extremity at 6 months* – In some cases, such as this one, particularly in younger patients, the use of DRG stimulation in CRPS can lead to a remission of symptoms. In those patients, they can often turn off the device and remain without disease signs or symptoms. In those cases, where they have weaned all medications and no longer need the device, we would recommend weaning the stimulation using it less each day until eventually the device is inactive. After 6 months with no recurrent symptoms, some patients elect to have their device removed.

Chapter 10

Programming concerns with spinal cord stimulation

Jianwen Wendy Gu

Principal Field Engineer, Boston Scientific Neuromodulation, Valencia, CA, United States

Scenario

A 50-year-old woman with failed back surgery syndrome (FBSS) has been identified as a spinal cord stimulation (SCS) candidate by her pain management physician and is undergoing a one week trial. She is met shortly before the trial procedure to confirm the pain areas that the physician has evaluated to be treatable with SCS, provide an overview of the procedure, set expectations, and answer questions. She says that she has pain in the low back and both legs including the feet, equal in intensity on both sides of the body. She rates her pain as an 8 out of 10 in the low back and 6 in the legs. The patient is informed that during the procedure, she will be awakened for "intraoperative testing" to ensure that stimulation is covering all of her pain areas or a reasonably large amount of the areas. She will feel a tingling sensation (paresthesia), and she needs to identify where she feels it. The goal is for her to feel the tingling sensation in all of her pain areas if possible. It is also explained to her that sometimes longer programming sessions may be needed to obtain complete coverage and that this may not be feasible in the operating room (OR) but a large proportion of the pain areas can be captured most of the time.

Variation: *Intraoperative testing is not performed*

The physician may choose not to do intraoperative testing and to place the leads over a specific vertebral region. In this case, the patient is informed that she may be asleep for the entire procedure or may not be asked for much feedback during the procedure.

She is informed that SCS does not usually provide 100% pain relief, although sometimes it does, and on average, patients experience 60%−80% pain relief. The patient is informed that after the procedure, she will be provided with a few initial programs to take home as well as details on what will happen during the trial week.

The Neuromodulation Casebook. https://doi.org/10.1016/B978-0-12-817002-1.00010-9
75

Within the OR, she is given quickly reversible sedation and local anesthetic. In this case, the physician chooses to use two 16-electrode percutaneous leads with 1-mm electrode spacing and a 67-mm electrode span to allow for assessing appropriate levels for coverage of the back and most of the lower extremities during testing in the trial. The 67-mm electrode span typically covers almost three full vertebral levels. The leads are placed paramedial and cover vertebral levels T8 to T10. Before intraoperative testing, impedances of the electrodes are checked, and they are all within the normal range. High-impedance electrodes may not be useable because of system constraints.

Variation: *Two electrodes in the middle of the right lead (electrodes 8 and 9 counting from the distal end of the lead) have high impedance*

The physician wipes the proximal end of the lead that inserts into the cable that connects to the external trial stimulator to clear any material that may be causing the high impedance. Impedances are now normal.

Variation: *Impedances are still high*

The physician moves the right lead in the rostral direction by two electrodes. The impedances move to electrodes 10 and 11, indicating an anatomical issue such as an air bubble. This usually clears after the procedure. During intraoperative testing, these electrodes are not used for programming.

Variation: *Impedances are still high*

The impedances stay high on electrodes 8 and 9, indicating a mechanical issue. The cable that connects the lead to the external trial stimulator is replaced, and the impedances are normal.

Variation: *Impedances are still high*

A different external trial stimulator is used, and the impedances are normal.

Variation: *Impedances are still high*

The lead itself is then replaced, and the impedances are found to be normal. During intraoperative testing, a pulse width of 200 μs and a frequency of 40 Hz are used initially. Because the physiological midline does not always coincide with the anatomical midline, stimulating with each lead separately is helpful for identifying the physiological midline and ensuring that stimulation provides bilateral coverage. Stimulation is delivered from the left lead in the middle of T9, and the patient is asked where she feels it. She says she feels it in her left thigh. Stimulation is delivered from the right lead in the middle of T9, and she feels it in her right thigh, confirming that the physiological midline coincides with the anatomical midline and the patient has bilateral coverage.

Variation: *Physiological midline is offset from anatomical midline*

The patient feels abdominal stimulation when the left lead is stimulated and left-sided stimulation when the right lead is stimulated, indicating that the physiological midline is toward the right of the anatomical midline. The physician is informed and attempts to reposition the leads by moving the left lead to the right of the right lead, thus trying to more closely bracket the midline. The leads now provide bilateral stimulation.

Amplitude is slowly increased until the intensity is strong but not uncomfortable for the patient. She is asked several times during this process where she feels the stimulation in her body and whether the amplitude is high enough or too high. Stimulation often can be moved to the midline by using equal amounts of current on each lead, which stimulates the space between the leads, and the patient reports feeling stimulation equally in both thighs. As the patient is assessed further, she is asked where she feels stimulation, trying to gain coverage all the way into her lower legs and feet, if possible, as well as the low back. Attempts to capture the low back usually require moving the stimulation in the rostral direction. At the caudal aspect of T8, she feels stimulation in her low back. The patient is asked if the stimulation is covering all of her low back pain, and she confirms that it does. Stimulation is then moved in the caudal direction. Again, the patient is frequently asked where she feels stimulation. At the rostral aspect of T10, she feels stimulation in both legs but not in the feet. The pulse width is increased to 250 µs to capture more dorsal column fibers, and she now reports feeling stimulation in the feet. The patient confirms that stimulation is covering all of her leg pain.

Variation: *Coverage of low back is only at the very top of the lead*

The patient has low back pain coverage when electrodes 1 and 2 are being used. Because lead migration is possible, especially during a trial when the leads are not usually anchored in place, it is more ideal if the physician moves the lead in the rostral direction by approximately two electrodes.

Variation: *The patient has focal pain*

If the patient has focal pain (e.g., only in the left knee), which often happens in complex regional pain syndrome (CRPS), the physician may elect to try to place the lead laterally on the side of pain to target the dorsal root fibers at the vertebral levels corresponding to the pain area (e.g., L4) and leverage their dermatomal selectivity. However, it should be noted that focal pain can still be selectively covered using dorsal column stimulation as long as the appropriate fibers can be selectively targeted.

Scenario (continued)

During postoperative programming, the same stimulation locations and settings that provided coverage during intraoperative testing may be tried, although often the electrical environment around the electrodes has changed slightly and field coverage within the dorsal columns may not be very similar to how they were assessed in the OR. Additionally, the spinal cord is likely in a different location when the patient is not lying prone, as she was during placement of the leads. This can change stimulation requirements significantly. When the caudal aspect of T8 is stimulated in this patient, she reports uncomfortable rib stimulation in addition to low back stimulation. The field shape (current distribution) is adjusted by moving the anodes closer to the cathodes, which decreases the "volume of activation," the three-dimensional space within which nerve fibers are being stimulated. Shaping the volume of activation in this manner avoids stimulating the dorsal root fibers that can cause rib stimulation at this vertebral level. Once this change is made, she now feels stimulation localized to her low back. When the rostral aspect of T10 is stimulated, the patient has good coverage of the legs and feet and no unwanted stimulation. Both locations are stimulated simultaneously, and all of her pain areas are covered without any extraneous stimulation. The frequency may be adjusted to affect the character of the stimulation sensation. In this case, when the frequency is increased from 40 Hz to 60 Hz to modulate the sensation of stimulation, she finds this sensation to be smoother and more comfortable. Frequency is increased further to 80 Hz, and the patient says she prefers the previous setting, so the frequency is set to 60 Hz.

Variation: *Intraoperative testing is not performed during the trial*

The same strategy described for intraoperative testing to cover all pain areas with stimulation may be followed. Pulse width is adjusted to capture varying dorsal column fibers in the volume of activation. Note that amplitude may be adjusted to accommodate changes in pulse width. The distance between cathodes and anodes is adjusted to modify the volume of activation: the greater the distance, the larger the volume of activation; although at times, the depth of stimulation may be decreased if the electrodes are far enough apart. Again, amplitude and frequency may be adjusted for character of the stimulation sensation and desired comfort.

Scenario (continued)

Patients have been found to have different preferences for SCS waveforms, some prefer paresthesia while others prefer subperception. Furthermore, while many patients respond to both paresthesia SCS and subperception SCS, some patients respond only to one mode of SCS. Therefore, testing both modes of SCS may be necessary to determine whether a patient responds to SCS.

Titrating stimulation parameters (testing different stimulation parameters) is important for optimizing therapy with subperception SCS. Due to the short duration of the trial, a thorough titration of stimulation parameters is typically not practical until after the permanent implant except in certain countries that require month-long trials. Therefore, the stimulation parameters used during the trial are generally fixed based on what has been effective in other patients, and titration occurs after the permanent implant.

Due to the present inability to predict whether a patient will respond to paresthesia SCS and/or subperception SCS, "combination" programs that allow the patient to use both at the same time may be available and offered, which may be particularly useful because subperception programs usually take longer to "wash in" than paresthesia programs (hours to days for subperception vs. seconds to minutes for paresthesia). In addition, combination programs may provide greater pain relief by engaging multiple mechanisms of action.

In this case, the patient is programmed with a combination program consisting of a paresthesia program and a subperception program. She is instructed to adjust the amplitude of the stimulation (only the paresthesia part of the program) to keep the sensation in a comfortable range as she may experience changes in intensity in different positions. The amplitude of the subperception part of the program may be locked so that it remains constant. She is also provided with additional programs including both paresthesia and subperception options in case they are needed. The patient is informed that she may experience postoperative pain for the first day or two and she needs to try to distinguish between the acute pain and her chronic pain to evaluate whether SCS is effective. She is asked to evaluate improvements in function, not only pain relief. For example, can she walk, stand, or sit longer than usual? Functional improvements are often, but not always, correlated with pain scores and may also be important to evaluate. The patient is asked to refrain from movements (e.g., bending, lifting, twisting) that may cause lead migration but to try to do normal activities so that she can evaluate whether SCS is helping. She is again informed that she will be contacted throughout the trial to ask how she is doing with the new programs and she may try different programs depending on how she feels. The patient is contacted the next day and asked how she is doing. She reports experiencing 50% pain relief for her chronic pain, although she still has some postoperative pain. She likes the sensation of paresthesia, so she is instructed to continue using this program. On the following day, she says that she is now experiencing 80% pain relief (perhaps due to the effects of subperception SCS washing in), and she has been able to walk more than usual. She is doing very well, and she is asked to contact the clinical team if anything changes or if she has any questions or concerns.

<u>**Variation:**</u> *The patient does not like the sensation of paresthesia*

She is instructed to decrease the amplitude of the paresthesia part of the program until she no longer feels stimulation.

<u>**Variation:**</u> *The patient reports that she is not experiencing any pain relief, indicating she may not be a responder to paresthesia SCS*

Because the effects of subperception SCS may not have fully washed in, she is asked to continue using this program for another day. The following day, she says that she is experiencing 60% pain relief and she is able to stand for longer periods of time than usual.

<u>**Variation:**</u> *By the second day, the patient is still not experiencing any pain relief*

She is instructed to change to a different subperception program. She is informed that the effects could take one to two days to fully wash in. The next day, she is experiencing 40% pain relief, and the following day, she is experiencing 70% pain relief.

<u>**Variation:**</u> *On the fourth day of the trial, stimulation is no longer covering low back and thighs, only calves and feet, indicating the leads have migrated caudally*

An appointment is scheduled to reprogram her later that day, and the pain areas are recaptured with stimulation by using electrodes rostral to the original active electrodes.

<u>**Variation:**</u> *The leads migrate too far to recapture the pain areas with stimulation*

The clinical team and the patient evaluate whether the trial up to the point of lead migration was successful enough to continue to permanent implant.

Scenario (continued)

At the end of the trial, the clinical team and the patient evaluate the trial as successful and decide to proceed to permanent implant. The leads are removed, and the patient is given time to heal before the permanent implant.

The physician chooses to use the same type of leads for the permanent implant to ensure suitable coverage of pain areas. During the procedure, the physician attempts to place the leads as close as possible to their location during the trial to ensure that stimulation can be delivered to the same region of the spinal cord that was found to be effective during the trial. The physician opts to omit intraoperative testing because the appropriate lead positions are known from the trial.

Because the patient may not be entirely alert due to sedation for the procedure and is experiencing postoperative pain after the procedure, only a basic programming session is done. A family member is available to inform that a more thorough session will take place at the first postoperative visit. The patient is provided with a few programs to use until the postoperative visit: the combination program that was effective during the trial plus a paresthesia program and a subperception program.

At the postoperative visit, the patient reports that, similar to the trial, she is experiencing 80% pain relief with the combination program. Since she is doing very well, the focus may be on further optimizing the subperception part of the combination program. Multiple versions of the combination program may be created, leaving the paresthesia part the same. For the subperception part, stimulation location, pulse width, amplitude, and/or frequency may be varied. Because therapy optimization is still an open avenue of research and optimal stimulation location and settings vary across patients, a wide range of stimulation parameters may be explored. The patient may be instructed to use each program for two to four days to account for wash-in period variation and to enter therapy ratings on the remote control for the stimulator if available so that the optimal program can be identified.

Variation: *The patient prefers to feel stimulation when she is awake but not when she is sleeping*

She is instructed to decrease the amplitude of the paresthesia part of the program until she no longer feels it when she sleeps.

Variation: *The patient would like to use different paresthesia programs when she is doing different activities (e.g., walking, standing, sitting)*

She is provided with combination programs that use the original subperception program with different paresthesia programs. Alternatively, she can be offered several different paresthesia-based programs to try, often using differing frequencies and pulse widths.

Variation: *The patient prefers subperception SCS only*

She is provided with subperception programs that vary in stimulation location, pulse width, amplitude, and/or frequency. These programs are set to run automatically for two days each.

Variation: *The patient complains she has to recharge her stimulator too often*

Optimal programming includes optimizing energy use as well as pain relief. If the patient has a rechargeable system, then she will quickly be inconvenienced if she needs to recharge every other day or more because of energy usage. Alternatively, with a nonrechargeable system, the battery will deplete much sooner with some programs than others and the patient may find she needs to have a surgery much sooner than expected to replace it. Finding programming solutions that use the least energy is an important part of the programming repertoire. Strategies for reducing energy usage include decreasing frequency, pulse width, amplitude, and duty cycle.

Scenario (continued)

An appointment may be scheduled in a month or two after she has had time to use the programs. She is instructed to contact the clinical team if she has any questions or concerns. At the next follow-up visit, program usage and therapy rating data are reviewed to identify the optimal program of the ones she has used. With this program, the patient is experiencing 90% pain relief and is very satisfied with the therapy. She is grateful this therapy has been available and has improved her quality of life. At this point, no additional follow-up appointments are scheduled unless the patient experiences issues or requests reprogramming. The patient experiences long-term pain relief and does not contact the clinical team.

Variation: *A new pain has developed and is not covered by the stimulation*

The patient contacts the clinical team after six months to say that she has developed a new pain area in the middle of her back that is not being covered by stimulation. Pain changing location or the development of new pain areas is a common occurrence. An appointment is scheduled to reprogram her. The additional pain area is captured by using electrodes rostral to the ones that are stimulating her low back. This problem may require several reprogramming sessions, and patient expectations are tempered appropriately. In addition, many new pains are not neuropathic or chronic and may not respond to SCS therapy. Ideally, the physician always endeavors to explore the true etiology of the new condition.

Variation: *Stimulation is no longer covering pain areas after a fall*

The patient contacts the clinical team after a year to say that she has had a fall and she no longer feels stimulation in her pain areas, indicating that the leads have migrated, which is confirmed by an X-ray. The pain areas are recaptured by using electrodes rostral to the original active electrodes. In these cases, if recapturing the pain areas with paresthesia through reprogramming is not possible, new programs are created by modifying the subperception part of the program to vary stimulation location, pulse width, amplitude, and/or frequency because paresthesia coverage is not always necessary to achieve pain relief. If these are not effective, the physician may decide that the patient requires lead revision or additional lead(s).

Variation: *Stimulation is in the right area but no longer helps the pain*

After five years, the patient contacts the clinical team to report that she is not experiencing any pain relief although stimulation still covers her pain areas. One of the challenges of SCS is that some patients lose efficacy over time. The reasons are unclear and may be due to worsening of the pain condition, the patient no longer remembering the degree of pain before SCS, and/or the development of tolerance. To remind the patient of what her pain was before SCS, she may be instructed to turn stimulation off for a week to determine whether stimulation is having any effect. After a few days she says that she turned stimulation on again because her pain was worse without stimulation, but she is only experiencing 30% pain relief in contrast to the 90% pain relief she experienced when she was first implanted. An appointment for reprogramming is rescheduled. The patient's program usage and therapy rating data may be reviewed to identify programs that were effective during the initial optimization phase. She may be given these programs, and an appointment may be scheduled to evaluate the programs after she has used them. At the follow-up appointment, three programs that are providing 90% pain relief are identified, and they are set to be used for a week at a time in random order. The goal is to prevent or delay the development of tolerance.

Programming support is provided to the patient as needed. Because SCS research is a fast-growing field, and SCS technology is constantly being advanced, keeping abreast of research and technology is necessary to provide the best possible care for the patient.

Note: All SCS programming and patient interactions are conducted under the direction of a physician. The variations in this chapter were chosen to provide an overview of possible programming scenarios for educational purposes, in particular how to manage complex situations that are atypical but important to understand how to handle if they arise. SCS patients usually experience positive outcomes without these challenges. In addition, this chapter represents an example of an individual approach to programming (under the direction of a physician) and may not be identical to the approaches of others. This chapter is not intended to replace or supplement the Directions for Use.

Section II

Deep brain stimulation

Chapter 11

Subthalamic nucleus deep brain stimulation for Parkinson's disease

Alon Mogilner

Neurosurgery and Anesthesiology, NYU Langone Medical Center, New York, NY, United States

Scenario

A 65-year-old woman is referred to discuss deep brain stimulation surgery for Parkinson's disease (PD). She was diagnosed with PD at age 55, after presenting with dragging of the right side of her body. In retrospect, she reported some "slowness" of gait 2 years prior and had seen a number of physicians prior to the diagnosis being made. She does not report a history of tremor. After the diagnosis of PD was made, she was started on carbidopa-levodopa by her treating neurologist with excellent results for a number of years, but upon presentation, now her disease had progressed to the point where she was requiring dosing every 3 hours to maintain her best functional state. While at her best "on" condition, she reported that she felt "almost normal"—however, she reported inconsistency in medication effect, along with occasional dose failures and dyskinesias both at peak dose and end dose which could be quite violent.

> **Variation**: *What if the patient was tremor-predominant instead of an akinetic-rigid fluctuator?* – The patient described here represents a common phenotype of patients referred for surgery, i.e., a patient who continues to obtain good benefit from medication but now experiences motor fluctuations and side effects such as dyskinesias, limiting the amount of good "on" time. In these patients, the effect of deep brain stimulation (DBS) is usually as good, but not better than their best "on" condition, but will result in fewer fluctuations, fewer dyskinesias, and possibly a need for less medication intake. Another subset of patients is those who are tremor-predominant. In those patients, the effects of DBS on tremor can be better than the best medication outcome, and surgery is usually recommended earlier in the course of the disease.

The Neuromodulation Casebook. https://doi.org/10.1016/B978-0-12-817002-1.00011-0
87

Variation: *What if the patient had rapid-onset symmetric symptom-atology?* – Idiopathic PD is usually asymmetric, with an initial indolent course and a so-called "honeymoon period" over the first few years after diagnosis and pharmacologic therapy. A rapidly progressive, symmetric presentation raises concern for other conditions including but not limited to multiple systems atrophy and progressive supra-nuclear palsy, conditions for which DBS is not recommended as there is no evidence of long-term benefit.

Variation: *What if the patient is inadequately medicated or refusing to take appropriate dosing?* – All DBS candidates for PD should be evaluated by a neurologist specializing in movement disorders to confirm the diagnosis and whether adequate medication trials have been performed before considering surgery. If a patient has not had an adequate course of medical therapy, DBS is not usually recommended as they may respond quite well to pharmacologic therapy for many years. An exception to this may be the aforementioned tremor-predominant patient, where medication may not be as effective for tremors as DBS will be.

Scenario (continued)

The patient was examined off-medication and noted to have a Unified Parkinsons Disease Rating Scale (UPDRS-III) motor score of 40, with rigidity, bradykinesia, and tremor bilaterally, slightly worse on the left side of the body. On medication, the UPDRS score was noted to be 15, with frequent dyskinesias. UPDRS part 3 scores off-medication are frequently used as a criteria for candidacy of DBS. Many have used a score >30 as a cutoff for candidacy. However, a true tremor-predominant patient, with little rigidity or bradykinesia, may benefit significantly from DBS therapy despite having a UPDRS-III score less than 30. The patient was deemed an appropriate candidate for surgery and scheduled for a neuro-psychological examination, which showed no evidence of dementia. She was deemed an appropriate candidate for subthalamic nucleus (STN) DBS. A mag-netic resonance imaging (MRI) scan was performed, revealing no mass lesions and no significant abnormalities, and the targets and trajectories are planned a week before surgery. Medical clearance is performed, revealing no medical contraindications to proceeding. The patient is scheduled for simultaneous bilateral STN DBS, with the pulse generator to be placed 10 days later.

Variation: *The neuropsychological examination showed evidence of moderate cognitive deficits, more than that would be expected from someone of her age affected by PD* – While DBS can be highly effective in treating the motor symptoms of PD, there is evidence that in patients with preexisting cognitive deficits, DBS may worsen these deficits. In such a case, there are those who consider the GPi a more optimal target as evidence has suggested a lower rate of permanent cognitive deficits after globus pallidus pars intenra (GPi) DBS than STN.

Variation: *A recent MRI scan shows atrophy abnormalities* – An MRI scan showing significant cortical atrophy raises red flags, notwithstanding the results of the neuropsychological assessment. If the degree of atrophy is extreme, surgery may not be recommended, or, at the very least, the patient is informed that the surgery is associated with a higher risk of both cognitive decline and other complications such as a postoperative subdural hematoma.

Variation: *The patient is on chronic anticoagulation for atrial fibrillation* – The anticoagulation will have to be stopped prior to surgery for an appropriate time period to assure that they are not at risk of intracranial hemorrhage. Depending on the agent, this may be anywhere for 3–7 days, but generally it is recommended as 7 days to be sure. The exact time to restart anticoagulation after intracranial surgery is a matter of debate and depends on a number of factors including but not limited to the clinical indication for anticoagulation, but generally it is suggested to be 3 days minimum.

Variation: *Should the surgery be planned as simultaneous bilateral* versus *staged bilateral?* – In a patient with bilateral PD, the decision whether to perform a simultaneous bilateral procedure versus staging the sides is one without a clear answer. There are those who always stage the hemispheres, with some recent evidence suggesting that the accuracy of the second lead placement is less than that of the first. Others will base their decision on the age and cognitive status of the patient.

Variation: *What if the patient has a non−MRI-compatible cardiac pacemaker?* – In these cases, planning can only be done based on computed tomography (CT). While some may obtain the CT on the day of surgery, it is my practice to obtain a high-resolution non-contrast head CT prior to surgery to allow for the planning to be done prior as with MRI. In addition, as CT-based planning is all indirect targeting based on the anterior and posterior commissure, placement of a stereotactic frame may result in artifact which obscures these landmarks.

Scenario (continued)

The surgery is performed under monitored anesthesia care with intravenous sedation, using a combination of precedex (dexmedetomidine) and propofol.

The procedure is performed with the patient off Parkinson's medications. The patient is brought to the operating room where the headframe is applied, and a stereotactic CT is performed. The planning system is used to obtain the stereotactic coordinates of the STN, and the first side is performed. A single microelectrode recording (MER) is performed, yielding 5 mm of cells consistent with STN, with cells responsive to passive movement of the contralateral upper extremity. The DBS lead is placed. Intraoperative testing is performed using the lowest contact on the lead, yielding tremor abolition at 2.0 V, with contractions at 4.5 V. The lead is affixed using the supplied cap. The frame coordinates are set for the contralateral side, performed in a similar fashion.

Variation: *Preoperative MRI with MR-CT image fusion* versus *MRI day of surgery* – Many centers now perform MRI prior to surgery and fuse the MRI to a stereotactic CT obtained on the day of surgery. This allows for the planning to be done prior to surgery and to shorten the actual surgical time. When this is done, the MRI is frequently performed with anesthesia support so that patients with severe tremors or dyskinesias will remain still in the scanner.

Variation: *Sparse STN cells only are found in recordings* – If the MER results do not reveal a length of STN on the order of 4–5 mm, the next step would usually be to perform a second trajectory. The decision where to perform the second trajectory can be based on the intraoperative physiology, i.e., the starting point of STN activity, the effects of microstimulation or semi-microstimulation, or a combination. Additional trajectories are performed at offsets of 2 mm from the initial trajectory. Alternatively, multiple simultaneous trajectories are performed, and the optimal one is chosen.

Variation: *Tremor arrest is not obtained intraoperatively* – If tremor arrest is not obtained by increasing the voltage, additional contacts are tested. If none obtain tremor relief, the side effect profile should be analyzed, and the lead placed in a different location. Frequently, for severe tremor, placing the lead 2 mm posterior may effect tremor relief.

Variation: *The patient does not wish to proceed with the second side* – If the patient expresses an interest to stop the procedure after one side and cannot be convinced to continue, the second side can always be performed at a later date. In rare cases, a patient may elect to have only a single side implanted ultimately. This is obviously their choice, but they still may receive bilateral benefits in many cases.

Scenario (continued)

Postoperatively, a CT and MRI are performed, showing no hemorrhage and good lead placement. The patient spends one night in the hospital and is discharged the following day, returning 1–2 weeks later for the generator placement. The patient returns to the office for initial programming several weeks later with suture removal and obtains good symptomatic relief.

> **Variation**: *The patient has a more complicated postoperative course –* Occasionally, particularly in elderly patients and/or those with marginal neuropsychological profiles, patients may act confused, disinhibited, or inappropriate after STN lead placement. Assuming there is no focal lesion such as a hemorrhage, depending on the extent of the postoperative confusion, the patient is either observed an additional day in the hospital or discharged to home.

> **Variation**: *At the postoperative visit, there is serosanguinous drainage from the implantable pulse generator incision –* If there are signs/symptoms of obvious infection, this can be observed with dressing changes. A true purulent infection will necessitate removal of all the hardware, likely including the brain lead. In less-overt infectious circumstances, one might consider oral antibiotics followed if needed by a surgery to try to salvage the system but clean the area and get it to heal. Typically, the bur hole opening is scarred in enough to seal the access to the intracranial space, and meningitis is exceedingly rare. The ultimate salvage rate is likely only 30%–40%, although this may be worth considering in patients in whom additional surgery later to reimplant the lead would be overly burdensome or unlikely and in patients who are benefiting tremendously from the stimulation, even at an early stage.

> **Variation**: *Despite multiple programming sessions, the patient is not obtaining the expected symptomatic relief –* If so, one could consider a head CT to make sure the lead has not migrated, which, although very rare, can occur. If there has been significant lead migration, the lead will need revision, which most likely will need to be performed stereotactically.

> **Variation**: *The lead is well-placed, but the patient is not obtaining meaningful tremor relief –* In rare cases of severe tremor, we have added an ipsilateral Vim thalamic lead to an STN lead system. There is little issue of overlap in trajectories, but the scalp incisions may violate principles of blood supply if not planned carefully. Success with this approach has been obtained.

Chapter 12

Globus pallidus internus deep brain stimulation for dystonia

Alon Mogilner

Neurosurgery and Anesthesiology, NYU Langone Medical Center, New York, NY, United States

Scenario

A 14-year-old female is referred to discuss deep brain stimulation surgery for idiopathic generalized dystonia. She is of Ashkenazi Jewish descent and presented in early childhood with dystonic movements of one leg, which later generalized. Multiple oral medications as well as botox injections were tried with limited relief over a period of several years and with different physicians. Genetic analysis revealed the *Dopa-responsive dystonia* (DYT)-1 mutation.

> **Variation**: *Non-DYT1 dystonia and differential diagnosis* – A series of other genetic mutations, often labeled by numeric DYT designations 1−21 (though other nomenclatures exist), have been identified, and successful use of deep brain stimulation (DBS) for dystonias of differing etiologies has been reported in dystonias of varying genetic origin. *DYT5*, which presents with gait difficulty, is remarkably responsive to oral levodopa and should be ruled out prior to surgery as it often does well with medications alone.

> **Variation**: *Secondary dystonias* – While DBS is effective in most cases of idiopathic primary dystonia, it has been used in secondary dystonias of a variety of etiologies, including cerebral palsy as well as tardive dystonia, and can be considered in cases of secondary dystonia in patients refractory to medical therapy, although with the caveat that the results are less predictable than those for primary generalized dystonia.

Scenario (continued)

A physical examination was performed and was consistent with generalized dystonia with prominent involvement of the left foot which tended to invert and roll inward when walking. She also had axial involvement with a tendency

The Neuromodulation Casebook. https://doi.org/10.1016/B978-0-12-817002-1.00012-2

for the back to arch when performing a variety of tasks involving either arm and when walking. With her arms in sustention, there was clear pulling up of the left shoulder, with the arm tending to adduct with wrist flexion and flaring of the fingers that increased when reaching for objects. She had a slight tilt to her neck/head. The right body was less impacted although the axial involvement did spread to involve the right hand with certain movements. She had no clear contractures although it was difficult to get a full range of motion in her left foot. A magnetic resonance imaging (MRI) scan of the brain was also performed, revealing no mass lesions and no significant abnormalities.

> **Variation**: *Fixed versus mobile deformity* – In cases of severe dystonia, it can be difficult to determine whether a fixed contracture is present. As contractures represent musculoskeletal deformities, they will not improve with DBS. A preoperative assessment of the patient under general anesthesia may help distinguish between fixed dystonic posturing and a contracture. Our practice is to perform our preoperative planning MRIs under general anesthesia where one can presumably distinguish between the two.

> **Variation**: *Is a brain MRI needed to assess the patient for surgery?* – While MRI scans of patients with DYT1 and other primary generalized dystonia do not usually demonstrate structural abnormalities, scans of patients with secondary dystonia, in particular posttraumatic dystonia, may be abnormal. In particular, if significant abnormalities are noted in the globus pallidus, consideration may be made for an alternative target (subthalamic nucleus (STN)).

Scenario (continued)

The diagnosis of refractory DYT1 dystonia is confirmed and, after more discussion with the child and her parents based mostly in the high likelihood of symptom progression and also high likelihood of significant symptom relief with DBS, plans are made for globus pallidus pars interna (GPi) DBS, with the lead placement to be performed under conscious sedation with microelectrode recording (MER). It was clear that she would likely not tolerate being fully awake in this surgery, primarily because of her young age to begin with, but in addition, she seemed to be anxious in general with office visits and healthcare–related concerns overall. Beyond the age of roughly 18–20 years, most patients are able to consider awake DBS or at least getting through the surgery with minimal sedation for the frame and bur hole placements.

> **Variation**: *Severe dystonia in a patient unable to tolerate conscious sedation* – As for the other DBS targets, so-called "asleep DBS" using intraoperative imaging (MRI or computed tomography [CT]) is an

alternative to the traditional awake surgery with physiologic mapping. This is particularly appropriate in the pediatric age group, as well as in others with severe dystonia who cannot tolerate conscious sedation. It should be noted that, as in STN surgery, general anesthesia does not preclude MER and/or macrostimulation, as a variety of anesthetic techniques have been described, which allow for traditional MER. Lead placement using intraoperative MRI guidance (as opposed to CT), however, is not compatible with traditional MER and is performed using solely anatomic targeting.

Scenario (continued)

The surgery is performed under monitored anesthesia care with intravenous sedation, using a combination of precedex (dexmedetomidine) and propofol. The planning system is used to obtain the stereotactic coordinates of the GPi, and the first side is performed. A single MER is performed, yielding cells consistent with globus pallidus pars externa (GPe), followed by border cells, and approximately 5 mm of cells consistent with GPi, with cells responsive to passive movement of the contralateral upper extremity. At approximately 1 mm below the GPi base, responses to visual stimulation consistent with optic tract (OT) are noted by shining a brief flash of light in and out of the visual field of the patient and appreciating the faint whoosh of firing nearby that is picked up by the microelectrode.

> **Variation**: *GPi as target* **in** *Parkinson's disease versus dystonia* – The properties of the GPi in dystonic patients, in contrast to those with Parkinson's disease (PD), can be variable, with the traditional increased GPi activity in PD not routinely seen in patients with dystonia. This common finding may make discerning entrance and exit from GPi more difficult.

> **Variation**: *The OT is not able to be identified* – In general, identification of OT suggests that the lead, at the very least, is not too lateral. While others recommend final lead implantation slightly lateral to the tract where OT is obtained, we consider implanting a lead in a trajectory with OT as long as macrostimulation reveals a reasonable therapeutic window.

Scenario (continued)

The DBS lead is placed. Intraoperative testing is performed using the lowest contact on the lead, yielding transient phosphenes at 2.0 V, with contractions at

4.5 V. The lead is affixed using the supplied cap. The frame coordinates are set for the contralateral side, performed in a similar fashion.

> **Variation**: *The GPi side effect profile* – In contrast to the STN (see STN DBS for PD herein), the internal capsule is medial to the GPi, and thus, contractions at low voltage suggest that the lead is too medial and should be moved laterally. Transient phosphenes represent OT stimulation and are not infrequently seen at the most ventral contact. Of note, as patients with dystonia may require higher voltages than patients with PD, there are some groups who place the lead a bit lateral to an appropriate MER trajectory to allow for higher stimulation amplitudes.

> **Variation**: *Choice of generator* – We routinely place dual-channel rechargeable generators for dystonic children/young adults because of the expected need for many years of chronic therapy. In patients without appropriate support and who may lack the ability to recharge, primary cell implantable pulse generators (IPGs) may be recommended.

> **Variation**: *Postoperative course* – An occasional microlesion effect is noted with GPi DBS for dystonia, as for other conditions. The effect, typically seen as a therapeutic benefit, may linger and fade over hours or even weeks in unusual cases. Most often it lasts a few days if seen at all.

Scenario (continued)

Postoperatively, a CT or MRI is performed, showing no hemorrhage and good lead placement. The patient spends one night in the hospital and is discharged the following day, returning 1–2 weeks for the generator placement. The patient returns to the office for initial programming, and a monopolar review is performed, yielding a reasonable therapeutic window for both leads. Clinical benefit is not seen at the visit. Over the next few months, however, significant improvement in dystonia is noted with chronic stimulation. She is able to return to the regular school environment and coursework and experiences a relatively normal track into young adulthood.

> **Variation**: *There is a delay in onset of efficacy* – In contrast to Essential Tremor (ET) and most cases of PD, the effects of DBS for dystonia may be delayed, taking weeks or even months. As such, the programmer bases the initial settings on a combination of the intraoperative findings as well as the therapeutic windows noted at the monopolar review. The general tendency is to use the most ventral contact that is tolerated.

Variation: *The patient seems to require high stimulation settings to obtain efficacy* – The patient obtains excellent benefit, but at 1 year postoperatively, the IPG, a primary cell, is interrogated and noted to have only 6 months remaining on its duty cycle. As indicated previously, rechargeable generators are preferred for young patients with dystonia. If recharging is deemed too difficult, switching to lower frequency stimulation may increase IPG life and yield similar benefit.

Variation: *One year later, the patient is admitted urgently with severe dystonia, necessitating intubation and mechanical ventilation* – The so-called "dystonic storm" can be seen with acute withdrawal of DBS secondary to IPG depletion or lead malfunction and is considered a neurologic emergency. Urgent recognition of the condition and evaluation of the DBS system with timely replacement/repair is indicated. Often these patients are able to return to their beneficial baselines within hours to days of restarting the therapy.

Variation: *No significant improvement is noted with well-placed GPi DBS leads* – As indicated previously, STN is an alternative DBS target for dystonia and may be considered in patients with suboptimal outcome from GPi DBS. Once programming options have been exhausted and imaging is performed, which shows the leads are reasonably positioned, considering STN as a new target or additional target should be encouraged.

Chapter 13

Thalamic deep brain stimulation for essential tremor

Alon Mogilner

Neurosurgery and Anesthesiology, NYU Langone Medical Center, New York, NY, United States

Scenario

A 68-year-old male is referred to discuss deep brain stimulation surgery for his upper extremity tremor, which has been characterized as essential tremor, also known as benign familial tremor. He reports a long history of tremors which began in early adulthood and initially were not bothersome. As they worsened over the years, he was initially treated with propranolol which improved his tremors, although it had to be discontinued secondary to hypotension. Primidone was tried as well, with limited efficacy. Alcohol intake significantly reduced his tremors, and his family history was significant for multiple family members (brother, father, grandmother) with similar tremor, although each with varying severity. Physical examination was significant for a mild, barely noticeable right-sided rest tremor, which worsened dramatically with sustention and intention. Gait was mildly ataxic. Spiral drawing revealed significant difficulty with an almost unintelligible signature. The patient was not able to drink from a cup without spilling.

Variation: *Tremor differences in Essential Tremor (ET) versus mixed Essential-Tremor-Parkinsons Disease (ET-PD) phenotype* – A patient with a similar long history who also demonstrates bradykinesia and rigidity on examination may represent a mixed *ET-PD phenotype,* i.e., patients who develop other parkinsonian symptoms after a long history of only tremor. Similarly, while "classic" essential tremor is an intention tremor and as such can be contrasted with "classic" Parkinsonian tremor, severe essential tremor can assume a rest component, just as severe Parkinsonian tremor can assume an intention component. If a patient referred for deep brain stimulation (DBS) with ET is noted to have additional Parkinsonian symptoms, consideration for a different DBS target (subthalamic nucleus (STN) or globus pallidus pars interna (GPi)) should be given. Moreover, patients with tremor-predominant

The Neuromodulation Casebook. https://doi.org/10.1016/B978-0-12-817002-1.00013-4

Parkinson's disease (PD) may be referred with the incorrect diagnosis of ET, and as always, a careful history taking and physical examination should be performed, with a recommendation for an evaluation by a qualified movement disorder specialist.

<u>Variation</u>: *The patient has primary head and voice tremor* – While upper extremity tremor is the primary indication for surgical intervention, a proportion of patients will present with severe head tremor and/or voice tremor. Frequently, these patients have concomitant extremity tremor, but occasionally a patient with insignificant arm tremor, but with severe head/voice tremor, may be referred for intervention. If so, bilateral intervention would be indicated, as midline tremor is most responsive to bilateral intervention.

<u>Variation</u>: *The patient has tremor secondary to multiple sclerosis* – Patients with multiple sclerosis (MS) frequently suffer from severe intention tremor. DBS for MS tremor may afford functional improvement in a subset of these patients; however, many of these patients have a severe ataxic component to their tremor, and despite well-performed thalamic DBS surgery, improvement in overall function and quality of life may not be sustained. There is currently no clear indicator preoperatively if patients are more or less likely to respond to Vim DBS therapy.

Scenario (continued)

The patient was deemed an appropriate candidate for surgery and scheduled for a neuropsychological examination, which showed no evidence of dementia. He was deemed an appropriate candidate for DBS. A magnetic resonance imaging (MRI) scan was performed, revealing no mass lesions and no significant abnormalities. Medical clearance is performed, revealing no medical contraindications to proceeding. He is scheduled for left-hemisphere thalamic DBS lead placement. The patient undergoes an MRI prior to surgery, and the targets and trajectories are planned a week before surgery.

<u>Variation</u>: *An MRI showing significant cortical atrophy* – This MRI finding would raise a red flag. If the degree of atrophy is extreme, surgery may not be recommended, or, at the very least, the patient is informed that the surgery is associated with a higher risk of both cognitive decline and other complications such as a postoperative subdural hematoma.

<u>Variation</u>: *The patient is on chronic anticoagulation for atrial fibrillation* – The anticoagulation will have to be stopped prior to surgery for an appropriate time period to assure that they are not at risk

of intracranial hemorrhage. Depending on the agent, this may be anywhere for 3–7 days. The exact time to restart anticoagulation after intracranial surgery is a matter of debate and depends on a number of factors including but not limited to the clinical indication for anticoagulation. If, in fact, even temporary cessation of anticoagulation is deemed too risky, an alternative intervention such as gamma knife thalamotomy may be offered.

Variation: *The patient is not medically cleared for anesthesia* – While it is rare that a patient is so medically unstable that they cannot tolerate anesthesia for DBS surgery, if they are truly not surgical candidates, other treatment options include gamma thalamotomy or MRI-guided high-intensity focused ultrasound thalamotomy.

Variation: *Unilateral versus bilateral (simultaneous bilateral vs. staged bilateral surgery)* – Traditionally, thalamic DBS surgery was performed unilateral, targeting the dominant hand. Subsequently, bilateral surgery was performed in patients with significant bilateral symptomatology. Historically, unlike simultaneous bilateral STN surgery for PD', DBS for essential tremor was performed in a staged fashion. Recently, however, simultaneous bilateral thalamic DBS is being performed more frequently.

Variation: *Preoperative MRI with magnetic resonance-cat scan (MR-CT) image fusion, versus MRI day of surgery* – Many centers now perform MRI prior to surgery and fuse the MRI to a stereotactic computed tomography (CT) obtained on the day of surgery. This allows for the planning to be done prior to surgery and to shorten the actual surgical time. When this is done, the MRI is frequently performed with anesthesia support so that patients with severe tremors or dyskinesias will remain still in the scanner.

Variation: *The patient has a non–MRI-compatible cardiac pacemaker* – In these cases, planning can only be done on CT. Although some may obtain the CT the day of surgery, it is my practice to obtain a high-resolution noncontrast head CT prior to surgery to allow for the planning to be done prior as with MRI. In addition, as CT-based planning is all indirect targeting based on the anterior and posterior commissure, placement of a stereotactic frame may result in artifact which obscures these landmarks.

Scenario (continued)

The surgery is performed under monitored anesthesia care with intravenous sedation using a combination of precedex (dexmedetomidine) and propofol.

The patient is brought to the operating room where the headframe is applied, and a stereotactic CT is performed. The planning system is used to obtain the stereotactic coordinates of the thalamus. Microelectrode recording (MER) is performed to precisely localize the ventral intermediate nucleus (Vim) in the thalamus.

> **Variation**: *The debated use of MER in thalamic DBS* – While a variety of techniques have been described on using MER in thalamic DBS surgery, others routinely perform thalamic DBS surgery without MER, proceeding directly to macrostimulation through the DBS lead. Although targeting is often accurate enough for a macroeletrode placement on the first and only pass, if the electrode is slightly off target and adjustment needed, the larger track from the macroelectrode will have distorted the tissue and trajectory more significantly, possibly leading to erroneous recording of information if recording is subsequently used.

Scenario (continued)

The DBS lead is placed. Intraoperative testing is performed using the lowest contact on the lead, yielding paresthesias in the right hand beginning at a voltage of 1.5 V, which resolve after 5 seconds. Tremor arrest is noted at 2.0 V, with contractions at 4.5 V.

> **Variation**: *Paresthesias are noted but do not disappear* – Persistent paresthesias suggest that the lead is too posterior and thus close to the ventralis caudalis/ventral posterior medial (VPM) nucleus. If so, the lead should be withdrawn and placed anteriorly.

> **Variation**: *Tremor arrest is noted at 1.5 V, but contractions are noted at 2.0 V* – The small therapeutic window suggests that the lead is too lateral, close to the internal capsule. The lead should be placed medially and retested.

> **Variation**: *No paresthesias are noted, and no contractions are noted, with minimal tremor arrest* – In this case, the lack of side effects suggests that the lead is too anterior (and possibly too medial) and should be repositioned.

Scenario (continued)

Postoperatively, a CT or MRI is performed showing no hemorrhage and good lead placement. The patient spends one night in the hospital and is discharged the following day, returning 1−2 weeks for the generator placement.

Variation: *Same-day generator placement* – Some surgeons will place the generator the same day, under general anesthesia.

Scenario (continued)

The patient returns to the office for initial programming and obtains excellent symptomatic relief requiring minimal effort in programming.

Variation: *Programming is performed by an experienced neurologist, yielding good tremor relief. Over time, however, higher voltages are necessary, which result in mild dysarthria.* Some ET patients may require high voltages for tremor arrest, which can affect speech. If so, the newer DBS systems allow the patient to raise and lower the voltage as necessary. For example, the voltage may be raised while eating or writing and lowered while public speaking.

Variation: *Tremor control is initially excellent, but over time, the patient manifests breakthrough tremor.* A subset of patients with well-placed thalamic leads may have refractory tremor. Addition of a second, more anterior lead (Voa thalamus) has been reported, as well as the use of another target in the caudal zona incerta/posterior subthalamic region (CZi/PSA).

Chapter 14

Programming cases for DBS in Parkinson's Disease, dystonia, and tremor

Jay L. Shils

Anesthesiology, Intraoperative Neurophysiologic Monitoring, Rush University Medical Center, Chicago, IL, United States

DBS in STN for Parkinson's disease

Scenario

A 64-year-old male with an 11-year history of Parkinson's disease and bilateral segmental lead Subthalamic nucleus (STN) Deep Brain Stimulation (DBS) electrodes and implantable pulse generators (IPGs) placed 2 weeks prior to the visit presents to clinic for initial programming of his device. His primary symptoms were bradykinesias in both the upper and lower limbs. Up to the year prior to surgery, the patient was managed well medically and was on levodopa and an additional dopamine agonist. In the year prior to the DBS surgery, he had started to require more medication and was "on" for only about half hour at his peak medication dose. Surgery had been uneventful and bilateral STN leads were placed. The patient had no other comorbidities.

Initial programming of the device consists of trialing all ventral/dorsal electrodes in ring mode so that the stimulation field is cylindrical (the complete ring is stimulated) in a monopolar fashion. Initial stimulation parameters are set to a frequency of 180 Hz and a pulse width of 60 uSec. Stimulation is slowly raised until the patient feels adverse symptoms. In this case, he reports slight tingling in the opposite face, which resolves after a few seconds on one of the contralateral contacts. Other symptoms might include abnormal sensations such as persistent "tingling," abnormal motor sensations such as "pulling" or changes in speech such as "slurring" or reduced vocal volume. Increasing stimulation amplitude for him generally resulted in a reduction in the severity of his bradykinesia, although some contacts seemed to have a more immediate effect as active and passive limb motion and gait after most of the setting changes were tested. If no adverse events were appreciated, the

The Neuromodulation Casebook. https://doi.org/10.1016/B978-0-12-817002-1.00014-6

stimulation was raised to 4 mA. After testing all four ring mode contacts in this manner, the contact with the lowest amplitude that significantly reduced the bradykinesias, while also having the highest amplitude that induced adverse events, was used as an initial focus to start programming with going forward.

> **Variation**: *Adverse effects are at or below the level needed for beneficial effects* – The programming configuration to try is to move to a bipolar stimulation pattern. Start with the most ventral contact as the anode and the next dorsal ring as the cathode. Follow the trial parameters used during the initial programming mentioned previously. Test all bipolar ring configurations once again looking for the amplitudes of beneficial therapy and adverse effects. Use the configuration with the lowest amplitude needed for treatment of the tremor and the highest amplitude that begins to generate adverse effects to provide the largest therapeutic window.

> **Variation**: *Adverse effects are still at or below the beneficial effect for tremor in a bipolar configuration* – In this case, moving to the segmented sections of the rings is appropriate. All segments should be individually interrogated once again to locate the amplitudes of best therapeutic effect as well as the amplitudes of negative effect. Testing individual segmented contacts in a monopolar configuration should be tried first. If there is not enough beneficial effect for reducing the tremor, then moving to segments directly ventral and dorsal tied together is appropriate.

Scenario (continued)

A day after the initial programming, the patient returned to the clinic with significant cramping in his upper extremities. Multiple attempts to change the electrode contacts by moving to segmental contacts and changing to more dorsal contacts were performed, leaving several minutes at most settings to determine whether changes could be appreciated. Some configurations reduced the cramping but did little to improve the bradykinesias. Additionally, pulse width and frequency changes were tried with similar results. Often, a simple but disciplined adjustment can be performed by working through each contact change and adjusting frequency up and then down, pulse width up and then down, and raising amplitude for each change without an adverse effect. Simple testing of limb stiffness can easily be determined at each change as well as gait testing if results seem promising. This plan was followed generally in this patient, but none of the aforementioned programming modifications improved the cramping while not improving the bradykinesia. Such a stalemate within the universe of parameter changes is suggestive that a reasonable balance of benefits without adverse consequence is unlikely and the patient had to be taken back to the operating room to have the lead moved more laterally.

Once the lead was repositioned, the system was able to be programmed to relieve dyskinesias without causing adverse side effects. Over the course of 2 years, however, the patient noticed that improvements had significantly diminished in the lower extremities but were still present in his upper extremities. Additionally, the patient was noticing pain in one limb. It was assumed these were effects of loss of programming efficacy and/or progression of disease state. Multiple programming adjustments were again tried for him, as well as medication modifications. No improvements were seen in the lower extremities, and in fact, in some cases, there was also worsening of his upper extremities. Owing to the effect localized primarily in his lower extremities or back and, with deeper questioning, when he was upright but not sitting, the patient was sent for a lumbar computed tomography myelogram. The study demonstrated significant stenosis from L3 to L5. Program settings were returned to their original parameters, and he was sent to a surgeon to consider an operation for his stenosis and neurogenic claudication.

Programming for DBS in GPi for dystonia

Scenario

A young girl, aged 12, presented to the movement disorders clinic with a history since age seven of difficulty walking and then contractures and dystonic posturing. She comes in now in a wheelchair with significant dystonic posturing that is both painful and debilitating. The patient can no longer perform the basic activities of daily living. The patients' Burke, Fhan, Marsden dystonia scale (BFMD) score was 84. Testing shows that the patient was positive for the DYT-1 gene. She was a full-term, normal vaginal delivery and had what appeared to be relatively normal development until these symptoms began to appear. Medications had been of limited value over the intervening 5 years as her disability progressed to this point. The family moved to our area partly in an effort to seek a higher level of care for their daughter. At this time, she was essentially in "dystonic crisis," and an urgent assessment suggested that, given the overall state of the patient, bilateral GPi DBS was recommended. This was performed with her asleep in the Operating room (OR), intubated using a continuous infusion of propofol. During microelectrode recordings, anesthetic levels were tightly controlled to allow for higher levels of neural activity along the recording trajectory, but ultimately the data were useable but suboptimal.

Programming was initiated within hours of returning to the intensive care unit, with the patient still intubated and on a significant amount of propofol. The propofol infusion was reduced to allow symptoms to become apparent. All lead rings were individually tested at 130 Hz, 210 uSec up to 4 mA. The contacts with the least (no) adverse effects were chosen as the initial contacts. Specific adverse effects appreciated during this initial session were extra contractures and greater dyskinetic movements. Additionally, potential improvements in symptoms, such as reduced muscle stiffness, could at times be appreciated as well with careful observation and physical examination through passive limb motion. The initial parameters found in her were as follows: a frequency of 130 Hz, a pulse width of 210 uSec. The middle two complete rings of the lead were used because they demonstrated no adverse effects and a minor relaxation in her wrists. Initial intensity was set at the level where the patient seemed to indicate she felt some sensation in her arm and did not contort her face with potentially painful stimuli.

A few days after the initial programming session, the patient demonstrated a reduction of spasticity with some minor effects on contractions but no real change in lower limb posturing. The stimulation intensity was slowly raised by 1 mA, leaving the same contacts in place. No adverse effects were noted at this time. Over time, it was noted that the patient's posturing and dyskinetic movements had reduced to the point where she could be weaned off of the propofol and extubated.

> **Variation**: *In initial programming, the patient starts to exhibit minor transient contractions—"jerks"—with increasing stimulation intensity* – To differentiate direct stimulation of the motor system— medial electrode directly activating the internal capsule—or an actual effect on the GPi sensory motor region—which over time can prove beneficial—looking at how the response responds to different stimulation frequency is appropriate. Turning the stimulator to 1 or 2 Hz and increasing the intensity will have one of the two effects: (1) the muscle jerks will directly correlate to the stimulus pulses, demonstrating that the stimulus is directly driving the motor fibers of the internal capsule; (2) the muscle response will either not be present or may present uncorrelated to the stimulus pulses or may even become a continual contraction, demonstrating the response is due to a motor loop in the basal ganglia. Experience has shown that over time, this type of response correlated with a positive result.

Scenario (continued)

Over the next few months, the patient stopped showing any improvement in motor function. Increasing the stimulation amplitude and pulse width to try and the increase in the stimulation field did not improve the situation. At this point, the stimulation frequency was reduced to 60 Hz. Over the next 3 months, the patient was starting to be able to stand with minimal support. At this time, the most distal contact was added to increase the overall stimulation field.

Variation: *The patient returns soon after programming with new posturing of the arms* – In some case, when field size is enlarged, although benefit is seen immediately, in short order, it become clear the added stimulus recruits adverse effects. Often, shrinking the field by dropping out a contact or reducing pulse width can be helpful. Here, upon turning off the most distal contact, the posturing stopped over a few days. Additionally, investigating individual segments on the segmented lead may be beneficial in focusing the stimulation energy away from the internal capsule. Investigating each segment at times after the initial visit since many improvements may not appear during the initial visit, but rather only the loss of the adverse effect. Over time, adjustments can be slowly titrated to increase benefit.

Scenario (continued)

Over time, the patient slowly continued to improve. To see if we could increase the stimulation field to cover more of the GPi motor area, the most ventral contact was included. The addition of this contact caused "flashing spots" in the patient's vision. Upon turning this electrode off, the "flashing spots" stopped. Over the next 6 months, the patient continued to show significant improvement in reduction of abnormal motor movements and better control of motor function. The patient was able to push herself up from the wheelchair and stand steady with some support. Given the severity and duration of the disease, the patient showed bilateral plantar flexion contractions that were not being helped by the DBS. An orthopedician was consulted, and the patient had surgery to correct her ankles. Three months after surgery and rehabilitation, the patient was able to walk on her own.

Ventral intermediate nucleus (VIM) essential tremor programming case

Scenario

A 50-year-old man presented with a 25 year history of upper extremity tremor and a minor tremor in the lips. The patient is a lawyer, and his mouth tremor was starting to have a disabling effect on his ability to work. The patient's uncle and brother both have tremors of their upper extremities. Initially, the patient was able to "work around" the tremor. In the previous 2 years, however, it became more difficult to interact with other people and he became self-conscious of the tremors, causing him to minimize social interactions and put less than adequate time into work. The patient arrived to the clinic at the request of his wife. The patient was diagnosed fairly quickly with essential tremor and was considered a candidate for VIM DBS surgery. Initially a single-segmented lead was placed in the VIM on the contralateral side to the patient's dominant hand.

Seen approximately 2—3 weeks after the surgery, healing well, he was programmed first by trialing all ventral/dorsal electrodes in ring mode so that the stimulation field was cylindrical (the complete ring is stimulated) in a monopolar fashion. Initial stimulation parameters were set to a frequency of 130 Hz and a pulse width of 60 uSec. Stimulation was slowly raised until the patient felt adverse symptoms which included abnormal sensations "tingling," abnormal pulling motor sensations, and slurring changes in speech along with some hypophonia. Additionally, changes in tremor severity were also noted. In contacts where no adverse events were found, the stimulation was raised to a maximum of 4 mA. After testing all four ring mode contacts, the contact with the lowest amplitude that stopped the tremor, while also having the highest amplitude that induced adverse events, was chosen as the therapeutic contact to yield the largest therapeutic window.

Variation: *Dysarthria caused by the contact with the greatest tremor reduction* – Sometimes a notable side effect is brought out, although the benefit is also simultaneously highest. It was noticed at the initial programming session that the patient had some dysarthria in the most ventral contacts that were also the ones that offered the greatest tremor reduction at the lowest stimulator energy settings. Given that this effect can be transient in nature, the patient was sent home to observe this side effect over time. After about 1 month, the patient noted that his speech had slightly improved but not to the point where the patient could return to work. In cases where speech is not critical to the patient's livelihood, some dysarthria or hypophonia may be acceptable, but in this case, it was not. Individual ring segments were evaluated. By eliminating a single segment, the tremor in the mouth was completely resolved, and there were no speech-related adverse effects. This is where the most benefit of segmented leads can be found.

Variation: *Complete cessation of tremor in one area with some to minimal reduction in another area* – In patients who have multiple affected tremor sites, it may be best to find independent solutions for each and allow the patient the option of switching between them depending on their situation at the moment. This is especially necessary if adding multiple programs together starts to induce adverse effects. In this case, when using the settings mentioned previously, the tremor in the hand was reduced by about 60%. Other segments were added with increasing benefit to the hand but at the cost of increased speech-related adverse effects. A program was set up that used two segments on the same ring which would reduce the patient's hand tremor by about 90% while causing a small amount of speech-related adverse effects. A third programming option was also added that would reduce the hand tremor by 100% but cause moderate speech-related adverse effects. The patient could then choose the optimal programming option or a particular situation.

Scenario (continued)

After about 2 years, the overall effect of the stimulation started to diminish. A new evaluation of each contact was performed, and it was noted that tremor control required about 0.3 mA more amplitude. Another finding was that increasing the pulse width to 90 uSec also brought the tremor back under control. In this patient, the increase in pulse width reduced tremor without increasing the adverse effects. In talking with the patient, he found that he used the programming option that maximized his speech while in court, the programming option that maximized his speech and improved his hand tremor, but did not stop it completely, when meeting with clients, and the programming option that completely stopped his tremor but caused some speech issues during social gatherings or when eating. Increasing the amplitude on the complete ring program stopped the hand tremor but caused a little more dysarthria, while increasing the pulse width had the same effect with a much less negative change in the speech. While in the case of the segmental contact programs, the overall tremor reductions were found with both amplitude and pulse width changes without a significant difference between the two changes. In cases where different parameter changes generate similar clinical responses, the best option is the one that uses the minimum increase in energy output. Energy is defined as, *[pulse width]* × *[current²]* × *[frequency]*. For this particular case, an increase of 0.3 mA increased the energy output by 29%, while for the pulse width change to 90 uSec, the energy output increases by 50%. Thus, the amplitude was increased in this particular case.

After about 4 years, the stimulator effect was significantly diminished. The patient noticed that the effect of the stimulation stopped pretty suddenly. The overall "health" of the DBS system was interrogated. Battery output as well as electrode impedances were tested. The battery parameters were within their normal ranges, demonstrating the IPG output was functioning properly. In this case, it was noted that the impedance of the contact in use as well as to others had increased significantly, indicating a possible open circuit in those wires. After changing contacts to ones that had normal impedances and evaluating their effect, no significant improvement in tremor without causing significant adverse effects was noted.

On further questioning, he also conveyed that, sometimes, he felt almost as if there was a light or brief electrical shock feeling. He could not quite precisely place it on his neck or head. However, palpation along the wire along its path to the IPG in the chest area can reproduce the feeling, indicating possibly a wire or insulation break. In this case, the patient did not. If that sensation occurred with the sudden loss in efficacy, there could be a break in the insulation. In some cases, running an impedance test may not be able to locate an insulation break because there is still a current pathway from the tissue near the break to the case or other lead being tested. If the system has been in for a while (about a year or more) and the impedance has been stable, a large variation ($\sim 500\ \Omega$) from prior impedance values highly suggests change in the conduction pathway, and a short break needs to be considered. Imaging the lead pathway may be of help in these cases, but if it is a single wire, imaging will most likely not be able to visualize that change. If the break is in all leads, imaging could be of benefit. In either case, getting an X-ray is likely only of academic interest as the lead or extension will need to be replaced.

Using the aforementioned data, it was determined that there was a break in some of the lead wires inside the insulation. The patient was brought to surgery to have the system evaluated. The lead was disconnected at the extension/brain lead point, and the impedance of the brain lead was tested. In this case, the brain lead showed good impedances for all electrodes. The extension was then connected back to the brain, and the impedance was then checked through that pathway. In this case, the impedances of the electrodes found in the clinic were all high. It was determined that the extension wire was the problem and replaced. After surgery, the patient's DBS system was set back to the original settings with a good clinical outcome. In some cases, the wire breaks may be in the brain lead. In these cases, the brain lead needs to be replaced. The patient went on to continue using the device for many more years with continued functional benefit.

Section III

Peripheral nerve stimulation

Chapter 15

Case example and trial for peripheral nerve stimulation

Jeffrey E. Arle[1,2]

[1]*Neurosurgery, Harvard Medical School, Boston, MA, United States;* [2]*Neurosurgery, Beth Israel Deaconess Medical Center, Boston, MA, United States*

Scenario

A 32-year-old woman had been involved in a farm equipment accident several years earlier and required reconstruction of her right upper extremity with multiple skin grafts and muscle transfers. She had lost one and a half fingers on her hand and had a fused elbow but remained able to feel some areas in her hand and forearm as well as most regions more proximally. However, she also had likely avulsive injuries in several nerve roots in her neck on the same side and had had multilevel posterior decompressive laminectomies in her cervical spine. She had been treated since the accident for multiple areas of pain which included her right hand, or at least a significant but focal part of it, several parts of her forearm distally, and also headaches and upper neck pains on the right side using a combination of medications and a retrograde percutaneous spinal cord stimulation (SCS) electrode at the C1-2 region on the right. That electrode was located above the prior laminectomy levels. The implantable pulse generator (IPG) for it was located in the right subclavicular region. She explains that despite all of these attempts to treat her pain, she obtains only modest relief from the SCS and only a point or two on a good day in her hand and arm from the medication. She does relate that when she has missed doses of gabapentin, missing an entire day once, she again develops quite noticeably more severe pain.

> **Variation**: *The patient had no prior cervical surgery* – Given that this woman's pain syndrome involves one upper extremity and is fairly localized within the hand area, without prior posterior surgery to compromise the epidural space, it makes most sense to try a cervical SCS trial initially. In this case, the lead would best span the C5-7 levels. One can also consider adding peripheral stimulation to a previously placed SCS system (as in this case) to better cover the same area or

The Neuromodulation Casebook. https://doi.org/10.1016/B978-0-12-817002-1.00015-8

115

provide a slightly different type of waveform to broaden the ability of the therapy to benefit the pain.

Variation: *The patient had no prior accident or derangement of anatomy and no cervical spine etiology* – Although one would need to consider the basis for the pain syndrome at all in a case without prior trauma or obvious pathology in the cervical spine (e.g., nerve impingement or cord injury), it is sometimes reasonable to consider that the patient may have a severe peripheral neuropathy wherein, if the patient is not diabetic, the etiology is often difficult to discern. Such pain syndromes can often respond to stimulation therapy, in fact they are at their root "neuropathic" by definition and should at least receive an SCS trial. As noted previously, without a distortion of anatomy, the epidural space is undisturbed and a trial can likely be performed. One can also consider a noninvasive externalized trial as described here. In either case, the fact that there is no prior trauma or obvious anatomical etiology would not in and of itself rule out stimulation therapy.

Variation: *The patient had bilateral hand numbness and pain, worse at night, and focal periscapular pain only* – This combination of symptoms does suggest the possibility of carpal tunnel syndrome, and in some cases, especially involving the upper extremity because of the increased frequency of entrapment neuropathies, this etiology should be first evaluated using an adequate electromyogram-nerve conduction study (EMG-NCS). However, the periscapular pain can be considered either a cervical root injury potentially with a particularly focal aspect to it or a peripheral distal branch injury leaving the patient with pain, often related to movement of that area or palpation. An externalized trial can be tried in this region. Success can be fleeting and difficult to maintain in the long term with scenarios such as this however.

Scenario (continued)

Her goal is to treat the pain in her hand more effectively if possible. This has been the worst pain all along, and despite minimal benefit, she has continued to try to work in her family's flower and market stand, running the cash register, although any attempt to use the hand or arm for lifting or carrying is impossible. Her private and social life has been reduced to virtually nothing, and she has been suicidal at times because of the pain condition, fearing there is no hope and no solution to ease her pain.

On examination, her arm is clearly demarcated by the many mottled areas of healed skin graft and scars of prior incisions, as well as the distortion of anatomy from muscle atrophy and reattachment. It becomes clear that the most pronounced areas of pain are located in regions which can be delineated fairly clearly with a marker if necessary and follow regions innervated by branches of the median and ulnar nerves. She has had two prior nerve decompressions and feels, after further discussion, that maybe her pain worsened after the first of these and was not improved after the second. In any case, it appears that she may have the option of placing an electrode near sections of these nerves if they can be isolated adequately and then tunneling the leads up through the arm into the subclavicular region where her SCS IPG already is located. A merger of two 4-contact leads into a single 8-contact tail could be used to allow programming from her single IPG into both added peripheral leads and the already existing C1-2 lead, which occupies the other port. Another option could be to use an IPG which has 32 channels and four ports. Only one exists in the field currently, and the previously placed retrograde cervical lead may need to be replaced to be compatible with it if that option were chosen.

We discuss the possibility of doing a noninvasive trial or, if that fails or is indeterminate, an invasive lead trial. A noninvasive trial can be tried using adhesive, nonneedle electrodes connected to a suitable amplifier that can generate waveforms of similar amplitude, frequency, and pulse width in a bipolar configuration. An Ojemann Cortical Stimulator does provide that capability, and the leads are placed overlying the typical regions of her pain in her hand and also more proximally. A range of parameter changes is tried within the general universe of parameters that are typically used. Ramping up the amplitude slowly during this testing allows one to distinguish motor or pain or overly sensitive thresholds as necessary. Leaving the settings that may be helpful in place for perhaps 30'-60' at a time allows for adequate assessment of whether or not this therapy may be useful.

She relates that after a few adjustments, there is very clearly a diminution of her pain in her hand, almost the entire area of interest, but not quite all of it. It appears that if an electrode could be positioned along the course of the median nerve or distal branches, maybe just within the vicinity, parameters could be found that would block much if not all of her pain. She is convinced to move forward with this plan, and surgery is scheduled.

Variation: *Incomplete or transitory benefit is achieved in the clinical testing* – It is critical to honestly evaluate the robustness and resilience of the relatively short trial period in the clinic using external surface electrodes. One may perform such a trial more than once when results are questionable, and often this is reassuring to both the physician and patient. However, it is often the case that the patient wants the stimulation to work and may overemphasize the benefit to "convince" the physician to place the permanent device, as they also often believe it to be a "last resort" therapy. Sometimes in the clinic, the stimulation that did provide shorter term benefit (on the order of several minutes but not for a full hour or more), but is then lost, can be turned off and then restarted or cycled on and off. In some cases, this is effective but is overall less reassuring for a permanent solution. Typically, external surface electrode trials are successful within seconds to minutes of finding a reasonable setting. If the benefit never fully covers a region, it can be discussed with the patient, and sometimes it makes sense to place permanent leads anyway because they may have more success with cross talk and other patterns once in place. Transitory benefit in the clinic, however, is less convincing and probably will not lead to a successful outcome if implanted permanently.

Variation: *No change or annoying increases in pain are generated* – This variation is also seen on occasion and is a fairly clear sign that permanent placement is going to fail. Pain or annoying additional dysesthesias can be generated by stimulation over the appropriate areas, and other than adjusting the frequency, pulse width, and contact locations, there is no other reasonable solution when typical "paresthesia-generating stimulation parameters" are used. However, in some cases, high-frequency (e.g., 10 kHz) or subthreshold stimulation can still be quite effective for pain. The problem is that many waveform generators available for trying this noninvasive trial do not generate frequencies high enough, and it cannot be tried in this type of trial setting.

Variation: *Multiple areas requiring several leads are needed* – Using external surface electrodes for a trial, one can more easily test disparate regions and multiple regions. However, when multiple leads seem to be needed to benefit the patient most, one needs to be creative in how to employ multiple leads, numbers of contacts, lead configurations, mergers, splitters, tunneling trajectories, and IPG placements. Again only one system currently offers a 32-channel IPG with four ports. This system may be best suited for many of these more complex cases involving multiple leads and channels.

Variation: *The case may need to be handled with the help of a hand or plastic surgeon* – This example case involved the upper extremity and hand, in particular, and given derangements and the comfort level of the surgeon with this anatomy, it may make most sense to seek the aid of a hand or experienced plastic surgeon who can help with dissection. One critical feature to determine is whether the case requires dissection of a nerve itself or a field type of stimulation can be used to best effect. The former (as described further in this chapter) may require assistance.

<u>Variation</u>: *Her extremity had been amputated, and she had stump pain* – Stump pain, involving the tissue in the stump closure or margin, may be treated well either using SCS in traditional fashion or using peripheral nerve stimulation (PNS). It is most important to determine whether or not the most severe region of pain for the patient is fairly well localized or involves more of a larger region, especially if it migrates from one area at sometimes to another area at other times. The former is most amenable to PNS, and the latter may be amenable to SCS and potentially the need to provide additional PNS at a later point when it is clearer the SCS may not cover the pain as well in some regions, even if it is helpful overall. Most often, fairly focal stump pains can be trialed with surface electrodes externally in the clinic. As a contrast to this type of amputee pain, phantom limb pain is not likely to be benefited by PNS—but can be readily treated using SCS, and a trial should be arranged if the epidural space is accessible.

Scenario (continued)

The surgery in this case seems to be straightforward enough to contemplate without the help of a hand surgeon or plastic surgeon because the anatomy in the general vicinity of the pain is only partly distorted and she has compromised using her hand (from prior grafts and digit loss) to begin with. Two leads can be placed under the skin using an Epimed needle with plastic stylet allowing fairly direct positioning of the contacts within the region of pain and consistent with where pain relief was obtained in the clinical trial. The patient is brought into the operating room with the regions of pain interest marked on the hand in the holding area. A discussion occurs there just to make sure the regions are correctly identified. In many cases, just touching the regions on the skin to make the marks is more painful, and it is important to ask the patient about this and let them know what you need to do before touching them at all. This is also true of the trial adhesive electrodes, and I recalled from that experience that she was sensitive but tolerated it. In some cases, the patient may need to do it themselves or simply point to the areas and no marks can be made until the patient is asleep.

She is put under general anesthesia and intubated. The region of the entire hand, arm, and up into the subclavicular area is all prepped and draped into the field. After an appropriate time out and antibiotics given, an incision is made just proximal to the regions of interest in the wrist area such that the length of contacts will lie under the regions and into the area just proximal to them. They are planned so that they can also potentially cross talk to each other to work with a larger field if needed. In this case, such a configuration is possible, but it is not always possible. The depth of the incision is only made under the dermis, taking care not to damage potential structures deep to that. An opening

of sufficient dimension is made to allow for passing a tunneler more proximally in the arm and suturing a 2-0 silk anchoring stitch around the lead. The
Epimed needle is carefully passed to the target distance, the stylet removed,
the lead placed to the end, the lead stylet removed, and then the Epimed needle
itself removed, leaving the electrode in position. Immediately, an anchoring
suture or two is placed then to keep the lead in place there, a silicone anchor
can be used as well, but it adds bulk and may lead to a greater likelihood of an
erosion eventually. If possible, the wires are anchored slightly away from
being immediately under the incision.

Another small incision is made in the upper arm, just proximal to the
antecubital fossa and perhaps slightly medial over the biceps region; this
traverses the joint, and care should be taken in passing the tunneler at just the
right depth under the skin so as not to damage tendons or vessels along the
route. The leads can be anchored here again in a similar fashion. The subclavicular incision can then be reopened and the IPG removed. The tunneler
can then likely be made to traverse from the upper arm area to this incision,
taking care not to enter the axilla, the shoulder capsule, or deep within the
deltopectoral groove. The tunneler should be able to pass just under the skin
primarily the entire path. The leads can be thus brought to the subclavicular
area in this fashion and connected to the IPG, either by merging them as
mentioned earlier or using separate ports on the IPG. Extension wires may be
needed along the distance to allow for the entire passage. Excess wire can be
coiled deep to the IPG in the subclavicular pocket. I discussed the location of
the IPG with her ahead of time, and the need to maintain the cervical lead
passage to it. Another location for the IPG in a case involving the arm is in the
axilla region closer to the chest wall. The angle to place the tunneler from the
upper arm to this axillary location can be more difficult however. All incisions
were closed using subcutaneous Vicryl and nylon on the skin. Dressings were
placed. She was then awaked, extubated, and brought to the postanesthesia
care unit for further care and programming.

Variation: *The nerve isolated and lead sutured alongside nerve
itself* – Whether performed with dissecting assistance or alone, the
difficulty is in determining how to suture the typical cylindrical lead of
four or (more likely 8 or 16) contacts to the epineurium surface without
either making it too tight and compromising the nerve itself (and
potentially creating more pain) or securing it too loosely and allowing
the lead to migrate away from the nerve, rendering it less effective.
There is currently no electrode design that can wrap around the nerve
and safely "hug" its surface, requiring no suture, except the LivaNova
(formerly Cyberonics) vagus nerve stimulation lead—but this only has
two contacts, requiring all stimulation to be a bipolar configuration, and
the span along the nerve is typically 3 cm or less, which can be fairly
limiting when therapy is required to be maintained and potentially

reprogrammed over time. With a typical cylindrical lead, one suture of 2-0 or smaller Vicryl can be used to tie around the nerve and include the lead, just tightening it at the time to bring the lead into juxtaposition with the nerve surface, but no more. Then, several more ties are placed similarly, so that the lead is held just to the epineurium surface by three or four sutures. An anchoring suture then is placed around the lead more proximally and not involving the nerve, suturing the lead (with or without an explicit silicone anchor sleeve) to a firm but non-compromising part of tissue, such as an edge of fascia. Using an explicit silicone anchor is fine as well, but care should be taken to examine the thickness and robustness of the covering tissues and whether the additional material in the anchor may lead to erosion through skin at a later time. Finally, adding a strain relief loop after the anchor stitch, using perhaps two more sutures for the loop, is recommended if there is room. Over time, the fibrotic scar that forms over the lead and nerve in the area will hold the lead to the nerve well. If the lead is left with too much of a gap to the epineurium, more scar will form in between and potentially compromise the effectiveness of the electrical fields.

Variation: *Her pain is in the sural nerve region after 12 ankle surgeries and a fused ankle* – The case presented here involves the upper extremity only, but a similar situation could be imagined for lower extremity regions. Peripheral nerve pain in other areas and the approaches to rendering PNS for them will be detailed in several other chapters following. It is not too uncommon to see complex ankle fractures involving multiple repairs leading to complex pain syndromes. In cases where the epidural space cannot be accessed for SCS or if the region is quite severe and focused and unable to be adequately managed with only SCS, PNS should be considered and can be assessed with an in-clinic trial similarly. Internalized trial leads may be difficult in these cases because of the need to weight bear and move the foot region, unlike the hand which may be willingly immobilized for days at a time in a more conceited effort. Nonetheless, similar issues and concerns arise as in the upper extremity: Does a nerve itself need to be exposed or can field stimulation be used? How many leads are needed? Where can the leads be anchored and tunneled and where can the IPG be placed? (often in a thigh or buttock) And finally, will the leads reach the lengths needed or will extension wires be necessary, and if so, where will it be best to place the connections?

Scenario (continued)

After she was awake enough again, it became clear she had too much pain in the right upper extremity to clearly give feedback from programming. Ketorolac was given twice along with extra hydromorphone to come to terms with some of the pain, although this was obviously not able to eliminate the pain. Impedances were tested, and all reasonable and initial attempts to turn stimulation on revealed there were paresthesias generated within the painful areas more or less, but it was impossible to determine if pain relief was obtained. The stimulation was left on at a low level, and she was discharged home.

Follow-up occurred in the clinic approximately 12 days later when she returned for suture removal. Sites had healed well, and she had by then played around a little with amplitudes so she could report that there was some notable reduction of pain, but some edge of the area was not reached in her hand yet. We arranged for a reprogramming session, and after a couple of attempts, two separate programs were running simultaneously which managed to reduce the pain in all areas of concern between 40% and 80% depending on the regions. She became more functional in use of the arm and hand, which was still compromised in range of motion and strength, and able to continue working.

Chapter 16

Peripheral field stimulation for back pain

Jeffrey E. Arle[1,2]

[1]Neurosurgery, Harvard Medical School, Boston, MA, United States; [2]Neurosurgery, Beth Israel Deaconess Medical Center, Boston, MA, United States

Scenario

A 52-year-old man presented in clinic with a history of several back surgeries, ultimately manifesting in a fusion extending from L1-S1, including laminectomies at all levels from L1-L5. He noted that he had a disk surgery initially but then eventually two surgeries for stenosis several years later. Although there was some improvement at that juncture, he eventually had worsening of unremitting low back pain and eventually had a fusion of the entire lumbar spine in an attempt to treat this. Technically speaking, he had healed and recovered well, and the fusion construct looked reasonable on an magnetic resonance imaging scan and standing flexion extension films that had been performed. He had undergone several rounds of physical therapy without much benefit, and most recently, he felt that physical therapy (PT) made his symptoms worse. He also had several rounds of epidural steroid injections over the years, but none recently as other doctors had said they could not likely provide much benefit when access to the epidural space was now so compromised. Facet blocks and radiofrequency ablation also had been used in the past but after the fusion made little sense to contemplate. He also had tried acupuncture and massage, as well as what was now an escalating dose of narcotics, leaving him now on Suboxone. He also took a muscle relaxant and gabapentin, but felt these were of little benefit overall.

In locating and describing his pain, it was clear the pain was primarily near the midline scars he had but mostly just deep to the skin and fascia primarily overlying the paraspinal muscle regions. It was located in the lower lumbar area, just lateral at that region as well as near the posterior iliac crest margins. But it also extended higher up above L4—this is a region typically very difficult to cover ever with traditional spinal cord stimulation (SCS). Also, although there seemed to be a component of his pain related to stiffness or muscle spasm, it was more of a burning aching pain, and again muscle relaxants had minimal effect.

The Neuromodulation Casebook. https://doi.org/10.1016/B978-0-12-817002-1.00016-X

<u>Variation</u>: *The pain was instead right along the scar line and spinous processes* – This type of pain does not seem to respond well to peripheral field stimulation, but an external trial in the clinic could be tried using adhesive electrodes if the practitioner has the appropriate equipment. It may be worth trying lidocaine patches as well in some instances for this as the pain seems not to be related to adherence of scar tissue around the remaining bony margins in the midline and is not likely to be neuropathic enough. Placing electrodes along wither side (as in this case but perhaps closer to the midline) allows cross talk between them and may be able to provide some relief in the midline itself.

<u>Variation</u>: *The pain was only with movement and not there at rest at all* – This would be an unusual version of pain as it is almost entirely mechanically driven in nature and not likely to have much of a neuropathic component to it. Probably, there is an aspect of the prior fusion where a pseudoarthrosis may have occurred, and it may be difficult to see on any imaging, even computed tomography scanning. However, the use of PNS is not likely to be successful in such a situation.

<u>Variation</u>: *Pain was largely relieved when trying the TENS unit* – Although relatively uncommon in my experience, it is possible that a patient will get a fair amount of benefit from a TENS unit alone and can use it nearly all day long and does n't mind changing electrodes and managing the wires and so forth. Certainly, it avoids the need for a surgery and the associated risks. I advocate for getting as much out of nonsurgical means as possible, but most patients opt for an implanted device if given the option.

<u>Variation</u>: *No benefit using the TENS unit could be appreciated, and the feeling of the stimulation was largely annoying* – These findings when trying to maximize the use of the TENS unit make it quite un- likely that implanted field stimulation will be beneficial, but not impossible. In some cases, it is still worth considering trying an inva- sive PNS trial lead nonetheless. For example, if the patient is not very thin, it is much less likely that the information from a TENS unit will be helpful. There is too much soft tissue between the skin and fascia, and the attenuation of TENS unit amplitude requires far too much skin stimulation for patients to tolerate.

Scenario (continued)

With further discussion, it was apparent he had become somewhat knowl- edgeable about SCS but had been told by two other practitioners who placed trial leads that this would not be able to be done for him unfortunately because of the scar tissue throughout the upper part of the lumbar spine, severely limiting access to the epidural space safely. However, of note, during our conversation, he related offhand that he had "tried everything, including injections and those TENS units, and that thing you attach on your ankle". I asked him further about the TENS unit specifically—When had he tried it? Did he have his own? Did he only use it at physically therapy? Had he only tried it for 15′ or perhaps even 45′ at a time? When he had been using it? Did it help? If so, how much? Why did he stop using it if it helped? Why didn't he use it longer?

All these questions regarding a TENS unit lead in one particular direction in a case like this will unfold soon below. It turns out that he did not own one himself, but he had used it only in physical therapy several years ago and only for the time he was at therapy, which was about an hour once a week at that time. He said it had helped some, he thinks, while he used it, but did n't realize he could get his own or use it for longer periods of time. As he was trying to recall his experience with the device, he mentioned that it was as if he could n't turn it up high enough somehow. Maybe if he could increase the sensation it would be more helpful, that was a thought he said he had at the time, but then he had essentially abandoned it as another unhelpful adjunct to his care.

I mentioned to him that TENS units were now ubiquitously available and could be purchased easily and inexpensively online to try. But I mentioned one other aspect of trying a TENS unit more effectively—he could potentially get continuous, 24/7 benefit like a medication, with little or no side effects, and he would be the one controlling the therapy, where the electrodes were placed, and when he used it, as well as it would be invisible under his clothing. He was somewhat aware of these details, having used it at PT as mentioned, but had n't put it all together in quite this way. However, I then mentioned another aspect of this therapy, with a little bit of a teaser, based on the fact that he recalled the therapy previously had felt like "he couldn't turn it up high enough". I told him there may be another therapy that would work even better if this was still the case.

He agreed to have another go using a TENS unit more proactively and obtained his own for less than $50 online. I encouraged him to be very engaged with placing the electrodes and moving them into different configurations and trying multiple different parameter settings for extended periods of time as well. Most of these units turn off after short time periods (15′ or 1 hour, or some adjustable setting), but I also encouraged him to use the system as much of the day as possible by continuing to turn it back on or obtaining system initially that does not have an ever present automatic shutoff.

I also, however, explained to him that I did not necessarily think the device would help much or at all, although it could. I was more interested in what kind of feeling he had with the stimulation, what the quality of that feeling was (annoying, increased pain, pleasant, and so forth), and how high an amplitude he tolerated, or wished he could tolerate, to receive pain benefit, if any. I needed him to embrace the possibility of this therapy, offering information and potential benefit, but did n't want to raise his expectations inappropriately high.

He returned to the clinic in a couple of weeks and told me he thought there was some benefit to the TENS therapy and he tried to leave it on most of the day, although the unit he bought required him to restart it every cycle which was about an hour long. He was n't sure, however, if the feeling of the stimulation, a buzzy, massaging-like feeling, simply distracted him from the pain or was actually able to decrease the level of the pain itself. Either way, the stimulation feeling did not exacerbate the pain and was not annoying for him. He also did feel that if he could turn it up higher it would eliminate more of the pain—somehow it just did n't get deep enough. When he had tried to turn it up higher, at a certain point, the amount of stimulation was intolerable for him.

This was an encouraging sign for the simple reason that it implied that if the stimulation could be sourced from a deeper location (such as suprafascial, which is where peripheral field stimulation is placed) then it may be very effective for him. We agreed that the TENS unit had provided a reasonable benefit as a trial of sorts, but he was much more interested in having a permanent implant instead of continuing to use the TENS alone as a treatment, although I did mention that was an option for him. I mentioned that if the field stimulation we were contemplating was only partially effective (which is likely), then he could always supplement that with a TENS unit again. Surgery was discussed in terms of risks and benefits overall, including the possibility of infection, erosion, and lead migration—the main complications that can occur with these systems. It was scheduled for several weeks in the future on the elective schedule—of course, in the meantime, he could continue to use the TENS unit as desired.

In the preoperative holding area, the regions of most severe pain were outlined again with the patient sitting forward on the stretcher. The skin marking pen marks were meant to be useful for the actual surgical plan—the incisions for both lead access areas and the implantable pulse generator (IPG) pocket were clearly delineated. Four 8-contact electrodes and a single, 32-channel IPG were planned. The leads could be positioned so that they run vertically, parallel with the spine, using only two incisions located at the center of each side, passing the Epimed from the small incision superiorly, placing a lead and anchoring it, and then inferiorly and then anchoring it. The surgery is performed under general anesthesia, prone, but without the need for any monitoring. The leads are anchored initially to the suprafascial tissue with a 2-0 silk suture using a silicone anchor as the depth and overlying tissues are adequate, followed by a second anchoring suture encompassing both wires on the same side. Both sets of wires were then tunneled to one region on the right upper buttock where the IPG site was created. They were each placed dry into the ports of the 4-port 32-channel IPG, and impedances were checked. The impedances were all found to be within normal ranges, and the IPG was then placed with the excess wires coiled deep to it into the pocket. All three incisions were then closed with Vicryl and then Nylon for skin. Dressings were placed, and then he was turned back into the supine position, awakened, extubated, and brought to the postanesthesia care unit for further care and initial programming before discharge the same day.

Initially, as is often the case, there is too much discomfort in the immediate postoperative period to know whether or not the new system will be helpful or not. He is able to indicate that he feels stimulation generally in the right areas, and he seems to understand information given to him regarding recharging of the IPG and using his controller. Eventually that day, he is discharged, but he had turned the stimulation off before he left, according to his nurse who was with him for the 2−3 hours after the surgery. He returns in about 2 weeks for a first postoperative visit and suture removal. All of the incisions have healed well, and he has had some decline in the pain from the surgery itself. He has turned the system on but has not been able to fully evaluate how well it helps his pain yet. Fortunately, the representative from the company is present at this visit and goes through an hour of program settings which leaves him with four different programs to try over the ensuing weeks. Again, he reiterates that he does generally feel the stimulation in the areas of most pain, and it is not too dissimilar to when he used the TENS unit. I mentioned that this is good and he should be able to turn the amplitude up higher.

I asked about him several weeks later and the representative for the company tells me he has been doing well and uses usually one of two different programs and gets benefit. We call him directly and confirm that he is getting about 70% pain relief overall, and he is generally happy with this. It does n't take away all the pain, but it is a significant improvement where nothing else had been an option anymore.

Variation: *A lead migration seems to have occurred* – If the patient had good coverage and this was verifiable and reliable and then the patient returns at some future time saying they still feel some sort of stimulation but not near the same location they need it, it is more likely that the lead has moved. These leads are notorious for migrating despite having multiple anchoring sutures and strain relief loops. In some cases, the patient no longer even feels stimulation when it is "on"—the electric field is attenuated by the tissue and lack of nerve endings nearby. It is also not uncommon to have patients who think a lead has migrated when it has not. They are sure it must have moved because they fell or hit the area somehow and then their stimulation coverage or efficacy has been altered in some way. Naturally, this would seem to be a lead migration, but X-rays or surgical exploration may reveal the lead has not moved at all. Because it can be difficult to be sure of what has led to the loss of efficacy in some cases, it is important to relay this to the patient and begin a methodical evaluation. Usually, this involves obtaining X-rays of the area as a first step. If a lead has clearly moved on an X-ray, it is a solid justification for a surgical exploration and revision of the lead position. Intraoperatively, the system should always be electrically evaluated as well. However, often there is no obvious positional change on X-ray, yet the patient persists in claiming the system is not as helpful as it has been in the past. These situations are more challenging and may be manipulated by secondary gain concerns—or the claims may also be completely valid, and the electrical environment around the lead has changed in some manner, in which case the lead should be explored and revised. There is no perfect algorithm for resolving these situations in my view. Reprogramming should also be tried at least once, if not more, before surgical revision. The relationship and assessment of the patient along with the overall risk profile for a revision all need to be brought into the equation.

Variation: *Pain benefit wanes over time* – A similar, although more understandable, circumstance occurs when the patient describes that they have the same stimulation patterns and do get some relief, just not as much as they had in the past. Initially, the IPG needs to be evaluated and a discussion of recharging patterns and history should be obtained if the lead is a rechargeable lead. More likely, the explanation lies in the eventual maturation of scar tissue around the lead and in nearby tissues. Not only does it attenuate the general field amplitude reaching targeted nerve endings but it also alters the shape of the field, making it harder to continue capturing the appropriate efficacy. Reprogramming is almost always indicated, typically more than once, and a TENS trial may again be helpful, but ultimately a reexploration and revision is warranted. In these cases, paradoxically and unlike a claimed lead migration, there is more justification for a revision. However, in contradistinction, it is often that the therapy itself is not as effective anymore for the underlying pain. As opposed to a known lead migration where there was prior efficacy, a waning efficacy overall has a less sure outcome from a revision. In terms of the revision needed, painting with a broad brush, the lead can simply be freed from scar (if it is accessible along its length) or simply replaced in a nearby location, outside of its fibrotic sheath. The patient should be made aware that the same eventual development may occur again in the future in such cases.

Chapter 17

Peripheral field stimulation for Atypical face pain

Jeffrey E. Arle[1,2]
[1]*Neurosurgery, Harvard Medical School, Boston, MA, United States;* [2]*Neurosurgery, Beth Israel Deaconess Medical Center, Boston, MA, United States*

Scenario

A 37-year-old woman who had been in severe pain for over 4 years after three dental procedures was referred to the clinic. Two other oral surgeons and a pain physician who specializes in facial pain had seen her over the past 2 years and concluded that the pain was likely from damage to parts of the left superior and also, secondarily, to the inferior alveolar nerves. She had tried several blocks and high doses of gabapentin, carbamezapine, duloxetine, and pregabalin without noticeable benefit or at least any possibility to drop her pain level below about 8/10. Her pain was burning and aching with intermittent sharper searing pains, all consistent with damage to these trigeminal nerve branches. On the skin surface, the pain appeared to be delineated fairly locally over the upper and lower regions of the maxilla and mouth, extending almost to the tragus laterally and about 3 cm from the midline medially. Touching the area on the skin or using her teeth or mouth for chewing on that side, or sometimes on either side, exacerbated the pain, but at baseline, it was virtually always at least a 7/10. She had no particular loss of sensation in the area, although it was difficult to be sure there was no loss at all as it was intolerable for her to test the area well enough.

She also had managed to reduce her opioid requirements over the last year as they were not particularly helpful for the pain, but she still was taking about 20 mg of oxycodone per day in 2—4 separate doses. She had been up to 80 mg per day at times in the past. She had more or less lost hope that there was anything that could be done for her pain and was reluctant to make the visit, but her pain physician had insisted that she go. She also saw a pain psychologist for the previous 3 years who encouraged her to make the visit as well. After getting through basics of the history and having her describe and show the regions where the pain was worse (superior over inferior), I explained that this was potentially treatable with stimulation and described

The Neuromodulation Casebook. https://doi.org/10.1016/B978-0-12-817002-1.00017-1

129

how the electrodes could be placed within the face under these regions safely and so they were not visible. Importantly, I described as well how a trial for this could be accomplished noninvasively. She decided to go forward and arranged time to meet again in the clinic for the trial.

Variation: *The patient had prior trigeminal nerve decompression surgery* – In this scenario, the patient has been presumed to have had a typical version of trigeminal neuralgia and had a standard decompression and placement of Teflon cushioning to prevent pulsation to the nerve. If the history is truly consistent with this etiology, then facial trigeminal branch stimulation, as described in this case, will not be helpful. The surgeon must determine as best as they can whether the patient is more likely a failed case of typical trigeminal neuralgia or an atypical case with more distal nerve branch damage. Although it is possible that one could have both, it is quite rare. Often, the noninvasive trial in the clinic can provide more reliable information for a judgment on this.

Variation: *The patient has only brief episodic, lancinating types of pain occurring perhaps once or only a few times a day* – Often, patients with typical or atypical facial pain will describe it as a long history of very brief episodic pains—severe in nature, but brief nonetheless. They may occur several times each day or less frequently, numbering perhaps several per week. In either of these scenarios, such episodic pains are not as well treated with trigeminal branch stimulation as more persisting continuous pain. It is possible that the constant stimulation may prevent some pains from occurring or make the pain less severe, but it is very difficult to get trial information beforehand because of the infrequent nature of the pains. Most of the time when I have been presented with such cases, I have discouraged the patient from considering this type of therapy.

Variation: *The patient has no discernible trigeminal branch distribution of their pain* – Atypical face pains take on many variations. If the pain, however, cannot more or less be ascribed to one or more of the typical trigeminal braches in location or etiology, then placement of electrodes optimally becomes quite difficult. The noninvasive trial in the clinic can be helpful in this regard, but many locations of the electrodes may need to be tried before determining whether permanent placement is warranted.

Scenario (continued)

She returned to the clinic, and I attached two adhesive electrodes to her face with the wires plugged into a connecting wire that plugged into the Ojemann

Cortical Stimulator. I had to add some tape gently over the electrodes to keep them in place because the adhesive was n'ot adequate. This was painful for her, but she tolerated it. She knew there was no other way to obtain the information. I placed one contact just lateral to the nasolabial fold superior to the upper lip and the other more lateral toward the anterior margin of the tragus. The field initially was too broad as I slowly raised the amplitude using approximately 60 Hz signal with a 100 µs of pulse width. I moved the electrodes slightly to narrow the field. No facial motor activity was observed or felt. Adjusting the frequency and pulse width periodically to see if there was much difference in coverage or efficacy eventually led to a very noticeable reduction in pain. She was both surprised and a little emotional as she said this was the first time in 4 years she had so little pain. I told her it was important to see whether it was robust and lasted. I left the settings and the room and went to see several other patients before returning to check on her.

When I returned, she noted that the pain had not gone away entirely and she could feel some pain a little bit more laterally than normal, but some of the other more medial regions were better than they have ever been. I adjusted the electrodes again and readjusted the amplitude, but left the frequency and pulse width the same. I also reversed the polarity a few times to see if that made any difference. The whole experience is a type of "seat-of-the-pants" process. In any case, when I returned, she said she did not want me to take the electrodes off. She said she had almost complete elimination of the pain and could touch and move her mouth more than ever before. She was very happy that there seemed to be hope. Eventually, of course, we had to remove the electrodes and she went home. I told her we could book the surgery or she could try this experience again if she was not convinced to go forward with a permanent implant yet. She said she had thought a lot about this and wanted to move forward with the permanent implants. I explained to her the risks of erosion and infection and not working and so forth. We scheduled the surgery several weeks into the future on the elective schedule.

Variation: *The patient is unable to tolerate external placement of the electrodes for trial* – One is faced with the consideration of placing the electrodes within the face directly as a trial. This can be done, but risks infecting the area or damaging tissues within the region of interest, yet is not out of the question when the patient presents with what otherwise appears to be a very good example of a treatable atypical face pain. The trial electrodes can be placed from the same region as where they would be put in permanently through a small incision and anchored there as well for the trial. Or one might consider trying to place them as if they will be permanent and then adding extension wires to them which are tunneled outside the skin nearby (perhaps a supraclavicular location). All of these variants are possible, but it is more invasive and has the risks mentioned previously. In addition, it is difficult to make enough room to include the extension connectors and excess lead wires under the skin in that region. Typically, the scalp is thin there and patients also have trouble tolerating all the hardware under the skin there for a week or so.

Variation: *The patient does not tolerate any stimulation during the trial—aggravates pain* – As in many versions of neuromodulation generally involving electrical stimulation, tolerance of the stimulation feeling itself is vitally important for the therapy to be deployed at all. In some atypical face pain cases, it is obvious when the patient's pain is made worse with stimulation. While always worth trying different frequencies, pulse widths, and amplitudes, these scenarios almost always eliminate this therapy from being considered.

Scenario (continued)

In the preoperative holding area, the regions of most severe pain were outlined again directly on the face of the patient using a skin marker. Care is taken not to cause too much discomfort for the patient when doing this. If the pain is severe enough, I let the patient mark the areas themselves. The skin marks are meant to be useful for the actual surgical plan. It is critical that the furthest distal extent of needed coverage is appreciated. Even small areas that ultimately are not included in the electrical field can be quite disabling for the patient and likely create a need to advance the leads if at all possible.

In this case, the plan was to use two 8-contact cylindrical leads, one spanning the superior alveolar region and the other spanning the inferior alveolar region. They would enter from an incision made just over the temporal region behind the hairline if possible and curved in such a way as to allow the tunneler to be passed from the posterior aspect of the incision to the subclavicular site for the implantable pulse generator (IPG). The subclavicular incision is marked as well in the holding area. The patient is given antibiotic coverage as appropriate and taken to the operating room where she is put under general anesthesia and intubated. Placed on a gel donut with the head slightly turned to the right, the incision region including the region for the IPG pocket is shaved, prepped, and draped into the field carefully. The ear is draped out of the field generally. After an appropriate timeout, the temporal region incision is made with a scalpel, taking care not to damage the underlying superficial temporal artery. The depth should extend until just shallow to these vascular structures in the area and shallow to the fascial plane there as well.

Hemostasis is obtained typically with a bipolar cautery, and the edges of the incision and soft tissue just underneath the skin is freed lightly to allow for suturing the anchor stitches and a strain-relief loop into the area for the two wires. Then the Epimed needle is bent in a gentle curve to recapitulate the contour of the projected path toward the electrode target in the face. It is passed gently staying superficial to any muscle layers but deep to the dermis. It is important not to have the lead, especially the most distal contacts, sitting too shallowly under the skin as this will assuredly lead to erosion eventually. The first lead is placed and, after removal of the stylets and needle itself, is anchored with a 2-0 silk suture only around the lead, without a separate silicone sleeve anchor. A separate distinct anchor can be used, but this increases the risk of erosion and incision breakdown in this area. The second lead is placed in a similar fashion. Typically, the overall distance from the medial edge of the incision that this method can accommodate is about 8 cm. The second lead is also secured with an anchor suture. Both wires are then brought into a small strain-relief loop and sutured again, if possible, within the region of this incision.

The subclavicular pocket is then created using a scalpel, followed by hemostasis and a combination of sharp and blunt dissection (a toothed forcep with a curved Mayo scissor is helpful generally, or just finger dissection may work). After packed with gauze, the tunneler is bent into an appropriate curve and tunneled between the two incisions, taking care to avoid deeper structures of both the neck and the periclavicular passage. The wires are then brought through the tunneler, connected under dry conditions to the IPG, and secured with the hex wrench. Then the IPG is placed into the pocket after removing the gauze, and hemostasis is obtained. Both incisions are then closed using interrupted Vicryl and a running nylon for skin. Dressings are placed, and the patient is then awakened, extubated, and brought to the postanesthetic care unit where she receives pain management if necessary (typically these procedures are not particularly painful) and initial programming of the leads. Often when the patient is awake enough, they can reliably report whether or not the stimulation is at least felt in the correct regions. Pain relief may not be readily discernible so early but may be within a few hours and after some of the initial trauma from the procedure has settled. Several programs are usually put into the system, and the patient is discharged home the same day. She seems to be particularly relieved but hopeful because she says she felt some changes already before discharge.

Variation: *The pain distribution is too medial to reach with electrode* – This is an important variation that ideally should be predicted before the patient is in the operating room. There are particular locations where it comes up more frequently—pain in the nasolabial region and pain almost all on one side of the face but with one region which has bilateral pain (e.g., supraorbital area). In each case, there is typically an area on the face in a natural skin crease (e.g., furrow between eyebrows) where a small incision can be made that will allow placement of a lead starting there to get to the more distal target and also allow for the tunneler to be placed from the temporal incision to it, thus allowing passage of the electrode wire all the way back to the same-side anchoring location and IPG placement. Without considering such measures, the target is either inaccessible or would require the lead be tunneled to the side opposite to the planned IPG.

<u>Variation</u>: *Too many electrodes are needed for one IPG or pain is bilateral* – As mentioned in the previous variation, bilateral targets are sometimes accessible from the same side with an intermediary incision. In most cases, however, if too many leads are needed or more extensive bilateral coverage is needed, then more creative planning and likely two IPGs are going to be required. Both can be placed in subclavicular pockets, and an additional IPG can even be brought down to the abdomen with extensions.

<u>Variation</u>: *The patient already has an IPG in the ipsilateral subclavicular region* – This is a rare situation but does occur with prior placement of neuromodulation therapies (deep brain stimulation (DBS) for example) or cardiac pacemakers and related types of hardware. Crossing the midline to place the peripheral nerve stimulation (PNS) IPG is less desirable than bringing the wires (typically with extensions) further inferior into the abdomen. It may also be possible, depending on the anatomy of the patient and the exact location of the previously placed hardware, to put the IPG in the axillary region near the chest wall.

Scenario (continued)

She calls once a day after discharge asking about showering and if she can remove the dressings in general; however, she appears to be doing ok so far, and she is next seen at her suture removal visit in about 12 days from the surgery. The company representative is there to meet her as well, and we find that one program she had in the postoperative area is working best, over the other 3, but is not covering the superior alveolar region well enough. Further adjustments of the electrode pattern are made, and a slight change in the amplitude range as well. She is sent out feeling more improvement. Overall, she states she is so thankful and that she has about 80% or more pain relief so far.

Although she was only seen one other time a few weeks after the sutures were removed, for a second programming adjustment, she calls about 18 months later saying she can feel the tip of an electrode. She thinks it has been nearly visible for a couple weeks but did n't call to mention this to us. Now it has been a day or so that she is pretty sure it is visible. When we look at it however, it is a very thin area of dermal coverage but not yet broken through. Perhaps it had been, but when we first encountered the area, it was still closed.

I strongly suggest we explore this in the operating room (OR) before it fully erodes and revise the depth and amount of tissue covering it so that it is still sterile technically. She agrees and we fit her into the OR schedule in 2 days, and under general anesthesia, I am able to use a #11 blade scalpel and make a horizontal cut of about 2 mm to remove the thin area of coverage and allow the lead pass to a deeper depth and suturing a 3-0 or 4-0 dissolving suture to bring more tissue over it. It is obviously critical not to damage the lead using the #11 blade, wires, or the outer insulating layer. The use of monopolar cautery is avoided because it almost always will spread to the contacts and cause excess heating and possible damage to the lead. Bipolar if needed should be adequate. The skin is closed over this with a subcutaneous suture of 4-0 Monocryl. She does well from this intervention and maintained her excellent pain relief without requiring further revision, and we have not heard from her again yet.

> **Variation**: *The patient returns with obvious infection at the anchor site* – Obvious infection with pus draining from the incision or a collection just beneath the skin but surrounding the hardware generally requires complete removal of all hardware, including any anchors, and ideally including any suture used to anchor the wires. It is possible to assess the result of the trial on its own merit if the infection is at the end of the trial period only and, if successful, return later when the course of antibiotics is complete and several weeks have passed without recurrence of the infection for placement of the permanent system. More problematic is if the just-placed permanent system is infected. This also unfortunately requires removal of the entire system. It can also be replaced similarly in the future.

> **Variation**: *The patient returns with complaints of lead being visible at the distal end through opening in skin* – With this description, she is told to come right in the same day to be seen in the clinic, where we notice that indeed one can readily see the first contact emerging from a small hole in the skin in the superior alveolar region. With a little manipulation (after applying isopropyl alcohol to the tip and some betadine), it is apparent that even up to almost three contacts may slide out of the opening. I explain that we have essentially two options: one is to go to the OR and remove the entire system on the premise that it is infected and needs to be removed to treat it ideally; the second option is to go to the OR, and considering that the contaminated area is small, is not draining pus, and has only been open for a couple days at most, revise the area as in the variation described previously. If the stimulation has been effective for the patient, they are typically quite reluctant to remove the system. The salvage rate with decontaminating and closing an eroded bit of hardware in these cases is over 60% in my limited experience. The area needs to be aggressively prepped preoperatively (the patient should apply bacitracin ointment to the area daily

until surgery—ideally within a week of being evaluated) and prior to the incision, with prepping performed in the typical sterile fashion but also by irrigating the area of tissue and the wire and contacts that are at all accessible from the outside (without moving the lead from position) with a combination of alcohol, betadine, and peroxide intraoperatively before closure. Depending on the type of erosion, one might consider the use of a vancomycin powder as well. Should the area break down a second time, it is still possible to perform the same type of salvage with positive results as well. The most important aspect of the salvage is making sure there is adequate perfused tissue coverage over the electrodes. This prevents reerosion and allows better healing in the meantime as well.

Chapter 18

Occipital nerve stimulation: trial and permanent

Jeffrey E. Arle[1,2]

[1]Neurosurgery, Harvard Medical School, Boston, MA, United States; [2]Neurosurgery, Beth Israel Deaconess Medical Center, Boston, MA, United States

Scenario

A 45-year-old woman presented in referral from a headache center with a long history of complex migraines and focal occipitoparietal head pain as well. The head pains seemed to be present all the time, or nearly so, while the more "typical" migraine, which was how she described it, came on every few days and lasted about a day or two. She had been prescribed numerous medications over the years, some of which may have helped at times but generally did not. Currently, she is taking a combination of tramadol and gabapentin. She had also had several injections to block the occipital nerve or trigger points in the back of her head. These had lasted for up to a few days at times. With further questioning about these injections, she did admit that they typically eliminated a significant amount of the pain for hours or a few days, but it always came back.

I asked her to show me where the primary area of pain was and whether she could outline it or delineate it well. She said it was always in the occipital area, spreading forward somewhat into the parietal margins. It began most often near what appeared to be the nuchal line. She did not believe she had had any history of trauma there, but couldn't rule it out either. After further discussion, it was more clear that this region of pain, more unrelenting, seemed nevertheless to bring about or have some role in initiating her migraine pains, which seemed also to originate on the same side of the head and extend in toward her eyes from that side.

I discussed how we could trial some stimulation in the area at the back of her head that might mitigate some of her pain, more likely having effect on the persistent occipitoparietal pain than the migrainous pain, which I was less sure about being related to or affected by stimulation. It seemed possible the migraine pains were a separate phenomenon and perhaps had been present in her life even before these other occipitoparietal pains. At this point, she said she would be likely to try almost anything and agreed to undergo a trial. I explained that noninvasive stimulation in the region was not really possible

The Neuromodulation Casebook. https://doi.org/10.1016/B978-0-12-817002-1.00018-3

because hair would prevent the electrodes from adhering to the skin. We would need to place the electrodes in almost the same way as when they are placed permanently: shaving hair, general anesthesia, small incision, and so forth. The wires would exit the skin for over a week, perhaps even up to 2 weeks some times to be sure of the results. I also explained that they often migrate out or otherwise become displaced, and getting a perfect trial to go well the entire time was not always the case. If successful, we could place the permanent system with the implantable pulse generator (IPG) several weeks later. Either way, the leads would be removed in the office once the trial period had ended. She agreed, and we planned the initial trial lead placement surgery date.

Variation: *The patient only has a migraine* – Many people have migraines and migraine variants. If the patient had predominantly a well-documented migraine syndrome only, it would not rule out using stimulation in this same way, but I would be more reticent to endorse a positive outcome. On the other hand, frankly, it is hard to be sure of any of the outcomes with these head pain and migraine variants in any of the cases. And, because most of these patients are desperate and will try almost anything that sounds reasonably safe and not likely to end in their death, it is almost immaterial how much I try to give a truly straightforward assessment of the likelihood of success. The outcome data from occipital nerve stimulation (ONS) for migraine per se have been mixed but anecdotally successful. Ultimately, it may be worth considering a subthreshold trial in most of these cases.

Variation: *The patient has clear occipital neuralgia (secondary to a trauma)* – Clinically likely true occipital neuralgia is a very good candidate for ONS, with or without successful occipital nerve (ON) blocks, especially suprathreshold stimulation using two leads per side and just below the subthreshold locations. All the same attention to details in terms of trial length, anchoring techniques, tunneling tech-niques, and diligence about preventing infection applies in the same way. The only difference really is in lead location and the physiological basis for this is unclear presently.

Variation: *The patient has other more vague complaints of head, neck, or trunk fibromyalgia-type symptoms* – Just as migraine is a softer indication to consider ONS, even subthreshold ONS, these other types of atypical head or body pains are also a softer call. The presumption is that pain mitigation occurs somehow in the spinal trigeminal nucleus where head pain is also processed. In some anecdotal cases, my own included, there is some kind of ability for this region also to alter pain from other regions. It makes more sense to be conservative in endorsing the benefit from ONS for these types of cases, but then the morbidity from a trial is very minimal, and in a 2-week period of time, the information can be gleaned, pushing the recommendation into the trial category most of the time. If the pain is entirely in the trunk and migrating fibromyalgia-like pains, it is quite hard to predict any benefit.

Scenario (continued)

The patient was seen and skin markings were made in the preoperative holding area, where an IV was started, the anesthesiologist saw her, and antibiotics were given. There are two types of occipital stimulation—suprathreshold and subthreshold—and the electrodes are placed in different locations for each. Typically, standard occipital neuralgia uses one or two leads placed horizontally across the region just above the nuchal line, capturing more or less both the greater and lesser occipital nerve distributions. However, more atypical and combination pains in this region seem to be also treated, perhaps better treated, by having the leads slightly more superiorly located, with the most superior lead in line more or less with the helix of the ear. This more superior location also seems to be more effective if the stimulation is also at a subthreshold amplitude. In this woman's case, I had decided to use the latter subthreshold plan, primarily because this did not seem to be standard occipital neuralgia and had this other (and possibly unrelated) migrainous component.

She is brought into the operating room (OR), put under general anesthesia, and turned into the prone position on a horseshoe head-holder attached to the Mayfield table attachment. We tuck her arms with cushioning to her sides and make sure her head is neutral or slightly flexed. I shaved her hair to approximately a centimeter or two above the planned superior location of the lead. I did ask her in the holding area whether she wanted me to shave bilaterally equally or she wanted to have an asymmetric haircut. She wanted it to be symmetric, so I extended the shave across both sides of the midline—similar in many ways to what one might shave for a posterior fossa craniectomy. The midline is marked for an incision approximately 2 cm long in a position where both leads can be placed in the correct location. After an appropriate timeout, this incision is made using a scalpel down to galea or pericranium—it is nearly the same in this region. Hemostasis is obtained using the bipolar cautery, and both leads are then placed by first bending an Epimed needle in a curve to match the planned trajectory and then passing it from the incision laterally to a target distance of approximately 6 cm from midline, taking care not to bring the needle too close to the undersurface of the skin. Shaving the entire extent of the trajectory and draping this into the field (although it seems unnecessary) is often recommended so that one can be sure they have not accidently pierced the skin with the Epimed needle, outside of the surgical field.

Once both leads are placed through the Epimed needle (as described elsewhere in this text—removing the stylet, placing the lead, removing the lead stylet wire, then removing the Epimed needle taking care not to pull the lead with it as it comes out), they are anchored just under the lateral margin of the incision with a 2-0 silk suture to the galea layer. Eight-contact leads are used. A second anchoring suture is placed around both leads together just at the incision area. I then use a tunneler and tunnel the leads from the incision to an exit inferiorly on the lower neck and slightly later to midline, opening the skin with a #15 or #11 blade over the tip of the tunneler. This avoids the midline area, which will be used for the permanent placement later if indicated. It is critical to bend the tunneler at a significantly angled curve to make the trajectory from the occipital incision across the occipitocervical junction appropriately.

The incision is closed with 2-0 Vicryl and a running unlocked nylon on the skin. Dressings are placed so that there is one at the exit site where there is often drainage, and clear adhesive dressings are placed around the entire region including the creation of a loop of the exited lead wires, bring them over the clavicle from back to front, and leaving the contact ends free. She is in modest discomfort but not unmanageable in the postanesthesia care unit but is able to be given a subthreshold setting after determining that she feels stimulation in the right area. We are able to discharge her home with some hydrocodone for the surgical pains.

She is called by the company representative twice over the following week, and things seem to be going fine so far. She believes the stimulation is having an effect as her overall pain has dropped down to perhaps a 5/10 generally, and she' has had no migrainous pains yet. I tell her in a separate call to try to engage in more normal activities if possible, to see how well she is managing the pain. I also explain that the pain benefit may not last, and extending the trial long enough is important. Also, she has really only received about 30% −40% change, so given the level of potential placebo effects in this type of case, I want to see more of a benefit if possible.

During the following week, we check the site directly 1 day in clinic to make sure it is not overtly infected. We also then plan for the removal visit at the end of the week. When she comes in that day, she explains that she gets actually about 70% benefit—her pain level baseline is down to about a two to three most of the time, and she has only had a brief and less intense migrainous episode. We gently but firmly pull the electrodes out, knowing that there are two sutures holding them in place under the incision. With a little force, the leads free up and are removed intact. She is very excited to get the permanent system placed and wants to schedule it as soon as possible. I explain that given the somewhat contaminated aspect of these trials, it is important to wait long enough for everything to heal well.

Variation: *The leads pull out or are otherwise displaced* – This is a common occurrence during ONS trials, unfortunately, the movement of the head and neck and the tenuous anchoring of the cylindrical leads make it difficult to keep them in place even for a couple of weeks. It is important to determine whether or not the patient had (1) benefit before the displacement of the lead(s) and (2) how many days they had the benefit for. If the benefit was reasonable (e.g., >50%) for more than 3 days, then it is not ideal but probably is adequate for placing a permanent system. In this case, however, as a subthreshold stimulation paradigm, it would be preferable to have benefit for at least a week.

Variation: *The patient has decent relief after only a few days* – As in the variation of lead displacement mentioned previously, too short of a trial is a reason to reperform the trial. It is unfortunate when the patient has been specifically referred for a permanent placement to have to tell them the trial needs to be redone. However, given the expense of the system and likelihood of revisions due to migration or IPG pocket revisions, it is worth as well as possible knowing up front if the therapy is truly likely to be beneficial for the patient. As such, even if the patient has relief of most of their pain within the first few days of the trial, it is ideal if the trial is continued at least a minimum of three full days for suprathreshold ONS and 7 days for subthreshold ONS.

Variation: *There is clear infection at externalization site of wires* – One needs to be vigilant and honest in the assessment of the exit sites at the end of (or during) the trial period.

Scenario (continued)

She returned for the permanent placement about 5 weeks later, most of her hair having grown back into place. No part of the incision looked problematic, however, and we proceeded to place the permanent leads and the IPG in the same exact manner as the trial except that instead of tunneling the lead wires out a short distance away along the posterior neck tissue, they were reanchored with a strain-relief loop as well at the base of the cervical spine in the midline, just superficial to the fascial layer. So, two anchors on the wires at the occipital incision are placed, then two more sutures on a loop at the base of the neck, before tunneling to the flank area where the IPG was placed. This method minimizes stress on the most mobile areas of these leads and shortens the overall wire length, which lessens the pull on the lead overall as well.

Variation: *Prior ONS leads had been in place and removed in the past* – If leads had been in place previously (e.g., prior leads had become broken or infected and then removed), it may be difficult to place new leads because of scarring within the galea and adherent to the skull. However, in general, new leads within previously accessed areas for ONS are still fairly straightforward, and typically the needle can directly penetrate and subdue any of the scar that has now matured in the area.

Variation: *A prior posterior fossa craniectomy had been performed* – This is an interesting variation that does not occur often. There can be pain from scar adherent to the dural patch or dura from a prior craniectomy, and there can be occipital neuralgia pain, either independently or as a result of the craniectomy surgery. In any case, it may be possible to place leads far lateral enough, or from a more lateral entrance point, to still obtain reasonable stimulation coverage. Careful planning and potentially obtaining a computed tomography scan preoperatively to see the exact demarcation of the bone removal can be helpful.

Variation: *The patient has had prior posterior cervical surgery* – Usually, unless the incision is in the low cervical area, prior midline cervical incisions would interfere with the anchoring of the ONS leads after tunneling from their occipital location. That' said, it is still possible to reopen a midline incision and dissect down to the fascial layer anyway and still anchor the leads there if desired. As such, ONS should not be ruled out.

Scenario (continued)

She had a postoperative course similar to when the original trial leads were placed. The IPG site was quite tender for her, however, and further medication with hydrocodone was needed for this. There was no particular difficulty placing the leads within essentially the same trajectories as for the trial. Some scar and granulation tissue was encountered while reopening the incision but did not cause any problems with retargeting the leads again. Eventually, over the next 3–4 weeks, she regained the same degree of benefit for her pain and had significantly minimized the severity and frequency of the migrainous activity, so there seemed to be some type of correlation between the two types of pain.

Variation: *The leads migrate* – This is frustrating for both surgeon and patient when coverage and benefit have been achieved and despite good anchoring technique, the leads move and coverage is lost. The patient will usually report that the benefit has waned or that the stimulation is clearly in a different location. In some cases, they will notice that the amplitude required has changed—likely going up. Any of these complaints requires an anteroposterior and lateral skull X-ray to see where the leads now terminate. Even the best anchored ONS leads can migrate in my experience. If a lead has been determined to have moved, it cannot be reprogrammed to reobtain coverage and benefit, it needs to be repositioned or replaced.

Chapter 19

Miscellaneous peripheral nerve stimulations

Jeffrey E. Arle[1,2]

[1]*Neurosurgery, Harvard Medical School, Boston, MA, United States;* [2]*Neurosurgery, Beth Israel Deaconess Medical Center, Boston, MA, United States*

Scenario

A 39-year-old man presented to the emergency room (ER) at his local hospital with obesity and severe localized pain in the right groin after doing some cleanup and sweeping in his garage the previous weekend. After examination by ER staff and a consultation from the general surgeon on call, it was identified that he likely had a nonincarcerated inguinal hernia defect. He was told to gently push in the region where there was a faintly noted bulge in the painful region and to try ice over the area or even lying flat with his head lower than his pelvis to try to reduce the hernia if the pain persisted at times. He was given some hydrocodone for more severe pain on an as-needed basis and told to make an appointment with the general surgeon as an outpatient in follow-up as he may want to consider surgical repair electively to prevent the possibility of an incarcerated bowel loop in the future, particularly if the pain only briefly subsides in the meantime.

Over the subsequent 2 weeks, he had little respite from the pain, which was worse with standing or bending forward and coughing. He required a refill of the hydrocodone from his primary care physician (PCP). As his appointment with the general surgeon neared, he began to have pain in the same area on the left side. Although not quite as severe or constant, it was nonetheless quite distracting as well, and he found himself spending much of his time out of work and lying flat in bed. Eventually, he was seen again by the general surgeon who felt there was a need to repair both hernias, one with mesh. He was scheduled for surgery within the next week where there was an opening in the schedule from a cancellation. Surgery seemed to go relatively smoothly, and both hernias were definitively repaired. Other than some brief urinary retention after the surgery which required an extra day with a catheter, he was discharged to home.

Unfortunately, over the next few weeks, he developed an infection in the mesh repair and a recurrence of the hernia on the opposite side repair requiring a revision. All told, he underwent three further surgeries in the inguinal regions

The Neuromodulation Casebook. https://doi.org/10.1016/B978-0-12-817002-1.00019-5

143

to address all the complicating developments, in addition to the original repair surgery, over the ensuing 8 weeks. After fully recovering and well healed, he returned for a follow-up with the general surgeon and complained that he had started to feel a burning type of pain, sometimes very severe, in the area of both groins, near but deep to his incisions. The pain was not quite the same in nature as the pain he had had when the hernias initially presented. Ice was somewhat helpful at times, but he could not maintain the application of ice all day long, and ibuprofen was ineffective.

His surgeon suggested that he may have some localized nerve damage from the multiple surgeries and scar in both areas. This was uncommon but not rare and often subsides over time or is treatable with antiinflammatories or gabapentin. He was prescribed gabapentin, initially 300 mg/night for 5 days, followed by 300 mg three times per day times 5 days, and then finally 600 mg three times per day. He tried this with ice and rest. Initially, he was groggy during the day with the gabapentin. This persisted for about 3 days but then subsided, and he had no side effects from the gabapentin. However, after taking the 1800 mg/day for a week, he realized there was little to no effect on his groin pains. In fact, during the month or so since the pain had begun, the pain had instead intensified and was unrelenting. He was referred to a pain specialist. Several injections that included bupivacaine and a steroid solution were tried on each side over a period of a couple months, with only 6–8 hours of benefit. He was told that the initial benefit was helpful in appreciating that there was likely a nerve tethering or entrapment from the prior surgeries in each region and that the benefit was from the bupivacaine, which of course then wore off within a matter of hours.

At this juncture, now many months out from the repair of the hernias and complications, and even longer since his original hernia presentation, he had been diagnosed with a neuropathic pain syndrome following from the bilateral hernia repairs and refractory to conservative care. His pain physician suggested that he try peripheral stimulation and a localized trial could be performed first to test whether this might alleviate some or all of the pain. The patient agreed, and an externalized trial was arranged in the clinic for roughly an hour of time. Sticky external electrodes were applied overlying the span of the area where the pain was primarily concentrated in each groin. The regions generally were areas of approximately 2 cm by about 4–5 cm, and a bipole was made and tested using the Ojemann Cortical Stimulator and an adapter for connecting these disposable leads. Within about 30 seconds, it was clear that a significant amount of the pain in each area could be eliminated and this relief maintained as long as the stimulation was "on". Two different frequencies seemed to be ideal for the two areas, and different amplitudes were found that were comfortable, but the patient was told he likely would benefit from implanted leads and an implantable pulse generator (IPG) for long-term therapy.

Variation: *The external stimulation trial is not helpful* – While the use of clinic-based, outpatient external stimulation is convenient and reasonably predictable for determining likely benefit from an implanted peripheral lead and IPG, it is sometimes difficult to obtain reliable results in the clinic. Leads may become detached frequently, and the patient needs to be checked on often enough so as not to waste time if the leads had come off and the patient thinks they are getting no relief. However, external stimulation is less helpful in some cases where an implanted system may still work well. Some would advocate placing the system even without a trial. There is no standard of care, although third-party payors may approve or decline to cover these procedures with or without a trial.

Variation: *External stimulation is helpful but by using a portable TENS unit instead* – An alternative to using a modified connector to the Ojemann Cortical Stimulator is simply to use a typical portable TENS unit. The main drawback of this approach is that the electrodes provided with most TENS units are much larger than is practical for many areas of the body, particularly on the face. Other regions, however, may allow them to work well, and this could be an alternative method for trialing. An additional advantage of using a TENS unit is that should the trial be successful, it provides a noninvasive therapy for the patient in and of itself. Perhaps the patient had n't tried a TENS unit previously, and this brief trial using one shows them how effective it can be. The damaged nerve or nerves are often not very deep below the surface of the skin, and the electrical fields can be adjusted adequately to achieve the same result in many cases.

Scenario (continued)

Approximately 3 weeks later, the patient was brought to the operating room and placed under general anesthesia after initially marking the regions of pain with a skin marker in the holding area preoperatively. This was performed with the patient in a supine position, as they would be on the operating room table to prevent positional distortion and maintain optimal lead placement. It's particularly important to map out the region of interest in a detailed way to best localize the electrodes. If the region extends longer than the overall coverage of the contacts on the lead themselves, one can try to use a longer contact span lead (although, in general, the longest span available from a given company is always used) or place two leads to ensure coverage. In this case, single 8-contact leads could be placed on each side with excellent field coverage of the pain. The exact nerve (femoral cutaneous, inguinal, and so forth) does not need to be isolated. The lead does not need to be juxtaposed exactly to the surface of the perineurium. In fact, the nerve itself often would not be dissectible and may be injured further if such dissection were tried in regions such as this, particularly when following prior surgeries or infections.

Variation: *The region of interest is not linear in shape or conducive to having a single lead cover it* – It's critical to map the region of pain in some detail prior to going into the operation room (OR), preferably while the patient is in a reasonable position so that it is not distorted too much by skin. If one eight-contact cylindrical lead cannot span the important length of the whole region of interest or there is too much width to cover with a single lead, then two should be placed, juxtaposed in such a way as to allow cross talk if needed between them and to span the whole area. Focusing only on a majority of the region instead of making the effort to cover it completely often leads to a failure of the therapy and a revision.

Scenario (continued)

The IPG was placed in the abdomen, and a four-port, 32-channel IPG was used in case further leads needed to be placed in the future. Dummy plugs were placed in the other two open ports. The abdomen made more sense given the inguinal location, and the added soft tissue in this patient allowed the opposite side lead to be tunneled safely in subcutaneous fat across the midline.

Variation: *Patient has an ostomy bag or scar from prior surgery* – An ostomy in the intended region for IPG placement is a contraindication for using that location for the pocket. Typically, it is too close to the ostomy or the adherence of the ostomy care and bag. Previous surgery scars are still possible to work around or even within, but avoiding the area is best if there is a reasonable alternative site such as the opposite side of the abdomen. Tunneling from front to back to use the upper buttock or flank tissue is an option under those circumstances.

Variation: *There is another IPG already in and on the opposite side of the body* – Trying to place two IPGs in separate pockets right next to each other, front or back, is not reasonable. There are likely to be difficulties charging and communicating with them. More than one IPG should be separated, such as both sides of the abdomen or both buttocks. IPGs should be a minimum of 8 inches apart from each other.

Variation: *The patient is very thin with little subcutaneous soft tissue* – Often the subclavicular region has little subcutaneous tissues, but it can be challenging in almost any of the common areas where IPGs are placed—abdomen, buttock, flank. The thinnest IPG should be used in these cases if possible, but the patient should be apprised of the increased risk for having an erosion or a painful IPG site. Rarely, however, is the lack of subcutaneous tissue a reason to avoid doing the surgery altogether.

Scenario (continued)

The patient recovered well and healed well. Programming initially involved using only simple bipoles on each side. Eventually, an additional anode was added on one side, and alterations in the frequency and pulse width were made on the other side. The patient had almost 100% pain relief and was very grateful. However, shortly afterward, he was making his way into a parking lot during an early winter cold-spell and partly frozen walkways and managed to take a very abrupt fall heavily onto his coccygeal area. Initially quite bruised around the area, he made it home and rested, using ibuprofen and ice over the next several days. His inguinal leads seemed to be working well after the fall. The bulk of the tenderness in the area subsided, but the region was discolored from bruising. In an unfortunate twist, however, he came to develop a severe version of coccydynia over the next few months. Any pressure or torsion to the area resulted in disabling pain, and he had trouble sitting anywhere or driving. A first injection into the area resulted in a few months of 50%−70% relief of pain. Subsequent injections, however, were short-lived or of negligible benefit.

Again working with the same pain physician and having no significant relief with subsequent injections locally, it was suggested he might try peripheral field stimulation in the coccygeal area. A single 8-contact percutaneous lead could be used and placed into one of the two free ports in the same IPG potentially or two leads could be used if needed. He was trialed in a similar fashion with skin electrodes and using the Ojemann stimulator. Benefit seemed to be possible in this trial, and he was eventually scheduled and brought to the OR, placed under general anesthesia, and had a single eight-contact lead spanning the pain region that had been mapped out again in the holding area preoperatively. The lead was anchored again securely with a silicone anchor and 2-0 silk suture. The lead was coiled and anchored in two further locations to the thin fascia locally with just a circumferential 2-0 silk suture as an extra strain relief in two places about 1−2 cm apart, and then the rest of the lead wire was coiled around within the soft tissues and the skin closed there. A dressing was placed, and the patient was repositioned laterally so that the side of the abdomen with the IPG was up. The patient was reprepped and draped in this lateral position, both incisions were opened carefully so as not to damage the wires or IPG, and the new coccyx electrode was tunneled to the abdominal pocket and connected to one of the two open ports of this four-port IPG after removing the dummy plug. The lead was secured using the microscrewdriver, and the IPG was then replaced back into the abdominal pocket. Both incisions were then closed and had dressings placed. The patient was then put into the supine position, awakened, extubated, and taken to the recovery room moving all four extremities for care and programming.

He recovered well from this coccygeal lead placement, and both the inguinal leads and the coccygeal lead are successful at eliminating over 80% of pain from those areas, giving him a much-improved quality of life. Programming for the coccyx pain requires three visits but ends with a large bipole spanning the entire length of the lead using two cathodes and two anodes. Unfortunately, a few months later, out shopping for groceries, he has his right ankle slammed by a shopping cart pushed by a small child not being watched closely enough by the child's mother. A relatively pointed edge of the cart's metal frame hits an area right over the sural nerve just above the lateral malleolus. He experiences a fair amount of pain initially and needs help getting home forcing him to find a friend to pick him up. Despite some initial benefits from ibuprofen and ice applied to the area, he eventually develops a significant neuropathic pain in this area from sural nerve damage. Again, higher doses of gabapentin are marginally helpful, and non-steroidal anti-inflammatories (NSAIDS) do not seem to be useful. He is generally unable to walk for even short distances as weight-bearing seems to enhance the pain, although there is significant pain even when sitting or lying down. Local injections are effective but again for only a few hours up to a day at a time.

Although initially inconceivable, a fourth peripheral nerve electrode is brought up as a possibility. If a single 8-contact lead could be used, which does seem to be possible, it might be tunneled also to the abdominal IPG site and placed into the fourth portal. The patient decides to give the trial a try, and a similar trial using skin electrodes is performed with very good results. The leads are positioned on the skin overlying the subjacent sural nerve area. After a few adjustments and adding tape over the leads to help them stick better, significant pain relief is appreciated that lasts well over 30 minutes in the clinic. When the stimulation is turned off, the pain eventually returns just as intensely within a few minutes. He is scheduled for surgery.

> <u>Variation</u>: *Spinal cord stimulation trial or dorsal root ganglion can be used if peripheral nerve stimulation fails* – Although peripheral nerve stimulation is an excellent and testable solution for many of these focal region neuropathic pain problems, they either are found not to work in a trial or fail at some point after implantation. The reasons for these failures are several including scar around the lead altering the electrical field and programmability, displacement of the lead, breakage of the lead, and a changed annoyance with the perception of the stimulation. Failure usually requires removing the system. Alternatives to this approach can include spinal cord stimulation, although this is often too imprecise or not adequate. It is also testable in a trial however. Also dorsal root ganglion or nerve root stimulation can be quite effective for these types of focal pain.

Scenario (continued)

On the day of surgery, the region of pain and allodynia is marked in the surgical holding area, and the patient is brought into the OR and put under general anesthesia in a supine position. In this case, the subject's painful lower leg is adjusted by bending the knee and positioning for better access to the sural nerve region using pillows. The trajectory for tunneling is planned by examining a way to tunnel from the sural nerve incision to an area just at the posterior inferior area of the popliteal fossa. In the subcutaneous tissues there, no nerve or vascular structures are at risk, and inferior to the leg bend it is a good area to resecure the lead to prevent migration overall during leg movement. A small incision is planned for the superior anterior thigh where, again, the lead can be tunneled and then reanchored just before going further superiorly across another motion segment at the groin.

Tunneling from the anterior thigh across the groin region is not without risk of damage to deeper neurovascular structures. And if there is little subcutaneous tissue to provide a buffer, then one should consider tunneling instead from the popliteal area to the buttock region and then from the buttock region to the abdomen where the IPG is located. In this patient, there is enough buffer, and the lead is brought to the abdominal pocket where it is placed into the fourth port on the single IPG after removing the last dummy plug. He recovers well from this fourth electrode placement, and all leads are programmed through the single IPG. He has significant pain relief in both groins, his coccyx, and one sural nerve. It is an unusual case in and of itself but highlights the ability to address multiple areas of miscellaneous peripheral nerve and regional pain with neuromodulation.

Vagus nerve stimulation and responsive neurostimulation

Chapter 20

New vagus nerve stimulation lead and implantable pulse generator placement

Jeffrey E. Arle[1,2]

[1]*Neurosurgery, Harvard Medical School, Boston, MA, United States;* [2]*Neurosurgery, Beth Israel Deaconess Medical Center, Boston, MA, United States*

Scenario

A 27-year-old man had suffered from epilepsy since the age of 8. He had two types of seizures ultimately, one with generalized tonic-clonic behavior and the other with staring and brief rhythmic chewing type movements followed by slight left arm twitching and abnormal posturing. For a few years in adolescence, he had no seizures, but by the time he was in college in his later teens, they recurred, forcing him to drop out of school and he was never able to finish his degree. Unfortunately, over the next 7 years to the present, he has had an escalating frequency of seizures, briefly decreasing in frequency with some of the medication alterations that have taken place many times over the years, only to resume their typical frequency after a few weeks or months on the new medication regime. At present, he has the generalized seizures every few months while the other type occurs a few times each week. On occasion, he can have even three in a day, while at times, he has managed for 2—3 weeks without any seizures. Maintaining employment with little training has been difficult. Relationships also have been difficult to maintain, and he has n't been able to drive in many years.

He has been brought into an epilepsy-monitoring unit several times over the years and found to have most often left temporal lobe onsets but with rapid spread to the right, and, on some occasions, onsets were thought to be from the right temporal region. He is right-handed, and discussion with him regarding intracranial monitoring and potential resection of brain tissue if a focus is found has left him feeling reticent about proceeding. He is reluctant to risk language function if the resection is from the left temporal area (which seems to be the most likely predominant focus location). Other options were discussed including vagus nerve stimulation, deep brain stimulation in the

The Neuromodulation Casebook. https://doi.org/10.1016/B978-0-12-817002-1.00020-1

153

anterior thalamus, and responsive neural stimulation placement (which would require intracranial monitoring). More discussions with the patient's parents and the patient himself took place, and eventually it was decided that he would try vagus nerve stimulation first. An appointment was made with the surgeon, and risks and potential benefits were described, including infection, displacement or breakage, hematoma, hoarseness during stimulation, damage to the carotid or jugular vessels or other neck tissues, and the beneficial possibility (especially over time) that seizures could be reduced in frequency and severity and may be eliminated entirely, although this was less than a 10% chance. On the other hand, he had been tried on over six different antiepileptic medications over the years, and a preponderance of data suggests that after only three different medications, his chances of being seizure free are less than 5% if further medications are tried. He agreed to move forward with the implantation, and surgery was scheduled.

> **Variation**: *The patient has focal motor seizures* – While there has not been identified any particular seizure type that responds best to vagus nerve stimulation per se, focal motor seizures are perhaps less likely to respond well. On the other hand, vagus nerve stimulation is not contraindicated, although identification of the focus is usually possible, and responsive neural stimulation may be better in this setting overall as the recording and stimulation electrodes can be usually placed very close to the focus and they can be positioned relatively noninvasively because the onset is often cortical.

> **Variation**: *The patient is physically and cognitively disabled with frequent generalized seizures and subsequent injuries and difficult to manage within their care facility at least partly related to the frequent seizures* – This type of situation is often one of the most rewarding and beneficial indications for vagus nerve stimulation. Getting the patient to agree to behave for the surgery workup, placement of the IV, and waiting in the preoperative area are all part of the challenge in these patients. Sometimes it is very easy and everything proceeds smoothly, and other times, it can be so difficult the surgery cannot be accomplished, although such cases are more rare. An important and often critical detail in getting such cases done expeditiously is having the legal guardian or power of attorney available at the time for consent by both surgery and anesthesia services. This should be arranged ahead of time and confirmed prior to the day of surgery.

Scenario (continued)

On the day of surgery, the patient was seen in the preoperative holding area and marked with a horizontal line in a skin crease in the middle of the left side of the neck and a similar length incision in the typical region of the sub-clavicular area, also on the left side. In the operating room, the patient was left in the standard supine position on a gel donut after the intubation. A laryngeal mask airway was able to be used as the head was only turned slightly to the right. Both incisions were prepped into the operative field, and the surgery began with the neck dissection after an appropriate time-out. A Debakey forcep and Metzenbaum scissors were used along with small blunt Weitlaner retractors to dissect through the platysma and then deeper, medial to the in-ternal jugular vein and lateral to the carotid artery, taking care not to damage the jugular wall or stretch it too much with a retractor. Often a retractor is needed to separate the carotid and jugular from each other, and one or two are needed more proximal in the wound opening at the skin margin and another potentially to retract the sternocleidomastoid muscle away for better vision. With further careful dissection, the vagus nerve is identified along the posterior lateral wall of the carotid artery, enveloped in a thin layer of connective tissue along the carotid. The nerve is able to be isolated and freed from this position for roughly 3 cm distance. The edge of the perineurium is able to be gently grasped by the forceps and tugged one way or the other to rotate the nerve or pull the nerve toward the surgeon to see and free up the connective tissue for 360° around the nerve. Full free dissection of the nerve is critical for placing the stimulating lead itself.

> **Variation**: *The vagus nerve is hard to see or is mistaken for the ansa cervicalis* – This possibility seems like it would be unusual, but it can happen more often than many think. The ansa cervicalis, particularly the superior root of this branched looping nerve, typically lies medial to the jugular vein as well, but there are two distinguishing features that make it distinguishable from the vagus nerve. First, it has a diameter almost always smaller than that of the vagus nerve. It is typically on the order of 1−2 mm in diameter, while the vagus nerve is typically be-tween 2 and 4 mm in diameter. Second, it lies medial but superficial to the jugular vein in the tissue planes of the neck. As dissection proceeds, it can be confusing, in particular, if the patient has had any prior surgery in the neck, that the vagus may not be immediately visible, even after finding the carotid and exploring the area of the carotid sheath. At some point, the ansa is apparent as a nerve structure somewhat in the right area. Both features described previously should raise a flag for the surgeon to continue looking diligently for the vagus nerve itself. A further point is that the vagus nerve can be very well hidden within the connective tissues along the carotid wall and even be located almost entirely behind or posterior to the carotid artery.

> **Variation**: *A small tear is caused in the wall of the carotid artery* – Despite the very brisk bleeding, the carotid is a muscular walled arterial

structure and, in general, very repairable with standard closures. A 6-0 Prolene, for example, can be used to efficiently close the opening if edges are reasonably approximatable. The opening can be closed even temporarily if the surgeon would be more comfortable waiting for a different surgeon who is more often operating on the carotid artery to help out rather than trying to hold pressure or clamping the carotid to minimize blood loss. The likelihood of incurring this injury is very low because of the robustness of the carotid wall, but in general, care should be taken obviously around the carotid and in not overretracting it, loosening internal plaque, or narrowing the lumen too much during the surgery.

Variation: *A small tear is caused in the wall of the jugular vein* – This injury is more worrisome in some ways than a carotid injury. If the tear is larger than a centimeter and irregular, it may not clot and heal over with simple pressure. It is important to realize that needles and catheters are placed into this vessel percutaneously all the time without massive hemorrhages into the neck. Most surgeons will have had passing experience with such procedures themselves, perhaps during their training. Holding pressure over the area, even on the skin surface, is typically adequate to stabilize the wall breach, while having a catheter filling the hole for the most part is also helpful in this regard. Intraoperative jugular wall damage should be considered similarly—if small enough, one can place a piece of tissue such as fat or muscle over the opening and then a pledget on top to be the target of the directed pressure against the opening. A piece of methylcellulose alone can be tried as well. Most of the time, 5—10 minutes of pressure is enough, but the area should be carefully inspected. Sewing anything to the wall of the jugular vein is not indicated as the suture likely will pull through and create a larger problem. It may be difficult to isolate the area, with the brisk bleeding and the need to avoid further injury to the wall by tearing the wall further. Having competent assistance for this may be needed. Asking for help from a second surgeon from vascular or the ENT service should be requested sooner rather than after a more disastrous situation has arisen. Holding pressure to fully occlude the jugular may be needed for periods of time and may not result in any increased intracranial pressure, but this is not ideal as a permanent solution.

Variation: *A large facial vein or branch is noted to be crossing the trajectory of dissection* – This occurs in fewer surgeries than anterior cervical spine approaches because the dissection through the neck for a vagus nerve stimulator lead placement tends to be more lateral than that in the approach for an anterior cervical fusion. However, it still occurs on occasion and may require that the large vein be ligated and cut, rather than simply retracted out of the way or coagulated. If the caliber of the vessel in the way is over 3 mm or so, it should be considered for ligation. A small amount of extra dissection often can be done to allow the vessel to be gently retracted out of the way in many cases. Care should be taken to look for and note other smaller veins that drain into the larger vessel. These can be easily torn in retracting the larger vessel.

Scenario (continued)

Once the nerve is free and visual access to the region is stabilized with the small Weitlaners, the lead is opened and brought into the field, making sure it is untethered on the drapes. The electrodes and anchoring spiral are brought down into the wound oriented so that the electrodes are placed on the nerve superiorly and the anchoring spiral placed inferiorly. The most superior electrode is placed around the nerve first, using a pair of narrow, toothless Cushing forceps. One can gently rotate the nerve along its axis and find the ends of the coil of the electrode to pull it from under the nerve or from its side, to place the coil around the nerve itself. The first coil is then gently slid superiorly to the extent it can be moved out of the way along the nerve, and attention is then brought to the second electrode coil. Placing the coils on the nerve proceeds until all three are on the nerve in order. Making sure diligent hemostasis is performed using the bipolar forceps as needed, the retractors are carefully removed so as not to dislodge the electrodes. A white silicone anchor that comes with the electrode kit is used just as the electrode wire emerges from the deeper region by sewing it to soft tissue locally, usually under an area of muscle. The wire is then brought around in a curve along the soft tissues, superficial to the sternocleidomastoid muscle but under the platysma margin. Two more 2-0 silk sutures are used to anchor the wire along this curve as further strain relief.

The subclavicular incision is now made using a scalpel, and then using a combination of the monopolar cautery and a curved Mayo scissor, the implantable pulse generator (IPG) pocket is made. Hemostasis is achieved using the bipolar forceps as needed, and the pocket is packed with gauze. The tunneler (using the sharper tip) is curved significantly and brought up from the subclavicular incision to the neck incision, taking care to traverse over the top of the clavicle. Also, the monopolar is used to breach the tougher connective tissue in the neck where the tip is attempting to come through so as to minimize tearing and torqueing on the tissues in the neck which may dislodge the lead or begin deeper bleeding. The leads are placed into the clear tunnel tube and brought through to the subclavicular incision. They are connected under dry conditions to the IPG (typically a new IPG which detects the heart rate and delivers a boost stimulus if the heart rate rises above a programmed threshold, often a harbinger of seizure onset). The IPG is placed into the pocket after removal of the gauze, checking the hemostasis again. The system is interrogated, and heart rate detection checked.

> <u>Variation</u>: *The lead impedances are not found to be within a normal range* – There is some basic troubleshooting that may be needed when testing the IPG and lead integrity at the end of the case. Often the sensing and communicating device is not communicating with the IPG and is typically out of position over the IPG. This is easy to correct. However, in situations where communication is functioning normally and the impedance values measured in the lead are either very high or low, one should consider reexploring the area of the electrode. Blood or irrigation fluid may have become sequestered in and around the contacts. Or the lead may have become displaced from the nerve entirely. This situation should be taken seriously, and the wound should be reopened if not resolved readily. In rare cases, if all other explanations are eliminated, the lead itself may be defective and need to be replaced.

Scenario (continued)

The incisions are then closed with subdermal suture and then with a subcutaneous monocryl suture for the skin closure. The system is retested. Once determined that the system is twice tested within appropriate operating range, the incisions are closed, dressings are placed, and the patient is awakened, extubated, and taken to the recovery area for further care. Most initial programming of the system occurs about 2 weeks later in the clinic of the treating neurologist.

Chapter 21

Revision or removal of vagus nerve stimulator lead and implantable pulse generator

Jeffrey E. Arle[1,2]

[1]*Neurosurgery, Harvard Medical School, Boston, MA, United States;* [2]*Neurosurgery, Beth Israel Deaconess Medical Center, Boston, MA, United States*

Scenario

A 38-year-old woman had been using a vagus nerve stimulator for over 10 years with fairly significant benefit, overall having only two seizures in the last 5 years that she was aware of. More recently, however, she had noticed that the small effect on her voice when the stimulator was activated had gone away. In fact, she wasn't sure the stimulator was even working anymore and wondered whether the implantable pulse generator (IPG) had run out. When she tried using the magnet, she had also noticed there was no sense that the stimulator activated. The woman made an appointment with her epileptologist, and the system was interrogated. The impedances all read as if there was an open circuit, suggesting that the lead wires had broken. An anteroposterior and lateral pair of X-rays was therefore ordered of the neck. On further questioning, she recalls falling down a few stairs several months earlier and landing awkwardly on her shoulder on the right side but twisting her neck over to the left. Except for some serious bruises, she broke no bones and recovered well over a few weeks and hadn't thought more about it until presently.

> **Variation**: *The impedance measurements are high but not showing an open circuit* – High impedances can also lead to ineffective stimulation, most likely by not recruiting the right fibers within the nerve because the amplitude of the stimulation is diminished when there is more scar tissue developing between the lead contacts and the nerve. Damage to the IPG could also result in high impedances, although this fault is much less likely and impedances rising due to scar tissue is very common. Often the patient will experience more seizures, and the question of a rundown IPG will

The Neuromodulation Casebook. https://doi.org/10.1016/B978-0-12-817002-1.00021-3

159

arise. However, it is far more likely to relate to diminished effective stimulation to the nerve and typically able to be programmed around by, for example, increasing the amplitude.

Scenario (continued)

The X-ray showed what appeared to be a very short gap in the wires from both contacts in the anterior neck area near where it was first anchored in the soft tissues. It was clear between this X-ray evidence and the open circuits when interrogating the system that the lead wires had broken. A new lead would need to be placed if she was going to continue obtaining the benefits of vagus nerve stimulation. This situation was presented to her in the office, along with the risk and potential circumstance wherein a new lead could not be placed, despite best surgical efforts, and she ended up ultimately having no stimulation. We also spoke about the possibility of placing a lead on the right vagus nerve, which she agreed to do if the lead could not be again placed on the left side. She was made aware that the improvement when using the right vagus nerve is typically not as high. She agreed, and we proceeded to schedule the surgery.

> **Variation**: *No lead malfunction is present, but the patient has a two-pin lead and now is desirous of the new one-pin implantable pulse generator –* This situation has recently become more pressing as the newer IPG has heart rate sensing and other features that make it clearly more effective than prior systems. There is, unfortunately, no adapter available to convert from one-pin leads to a two-pin IPG. In a general sense, there is not enough of an indication to justify the risk of removing the present lead and placing a new lead, with the risk that a new lead may not be possible to place. However, conceivably, the patient will agree to the risk in the hope of gaining the benefit, the patient has most of their seizures at night or cannot use their magnet readily (for a variety of reasons), or the patient has no warning for seizures and never is able to use their magnet effectively. These situations could push the justification more toward replacing the lead, but it should be emphasized that removal and replacement is never simple and has increased risk.

> **Variation**: *No obvious break is seen in the lead, and it is unclear whether the problem lies with the lead or the implantable pulse generator –* This is the trickiest situation to assess because it presently requires undergoing surgery and opening the IPG site, replacing the IPG with a new IPG, and then testing the system with the new IPG and the current lead. If the impedance readings are reasonable, then the problem is assumed to have been with the IPG, and the lead is left in place. However, if the problem persists, then the problem is likely to have been the lead itself. Now a new IPG has been already opened at this point. A new lead will need to be placed and connected to this IPG.

In the most extreme case with this problem, the lead is a dual-pin lead. As such, the new IPG opened to test will need to be a dual-pin IPG. But if the lead is found defective, not only does it need to be replaced but also it needs to be a newer single-pin lead, and therefore, another new IPG needs to be opened as well, which is a single-pin IPG. This problem could be readily resolved with a simple lead assessment cable that allows interrogation of the lead without using an IPG.

Scenario (continued)

On the day of the surgery, she was met in the preoperative holding area and had a mark placed with a skin marker near the prior neck incision and IPG incision. The neck incision had originally been made by a different surgeon many years earlier and had not been reopened since. The IPG incision also had been placed earlier and was more vertical and lateral than more standard recent IPG incisions, which are mostly horizontal and located a few fingerbreadths below the clavicle. The same incisions would be used again as they had healed well and did not bother her. She was taken into the operating room and intubated, left in the supine position on a gel donut. Enough of both sides of the neck were prepped into the field to place a right-sided lead if needed, although one could consider leaving the right side out and only redraping and prepping if the lead could not be replaced on the left.

After an appropriate time-out, the surgery begins by opening the horizontal neck incision and starting to look for the insulation of the lead coursing through the scar and soft tissue. Typically, it is evident fairly quickly, but in some cases, it requires some sharp and blunt dissection to begin to appreciate its course. Debakey forceps and Metzenbaum scissors are helpful to work through the scar tissue and start to follow the lead deeper toward the nerve. Very diligent and careful dissection can proceed with the further use of blunt small Weitlaner retractors, appropriately placed to avoid excess stress on possible connections with the jugular vein wall which is typically enveloped in scar that surrounds the nerve and the lead. Ideally, one can continue to follow the insulation of the wires as it leads to the nerve. A couple centimeters before reaching the nerve and the coils of lead that circumscribe it, there is a less thick, less insulated portion of the wires which can be helpful to know thus how far away from the coils one is. It is critical to stay as close to the insulation as possible most of the time. At certain points, it can be helpful to cut through connective restraints of scar or other tissue by using the monopolar cautery, typically on tissue right along the margin of the insulation. Cutting mode is more often best for this and should be applied in brief applications cautiously to mitigate heat buildup or spread which can damage tissue (such as nerve or blood vessel) unseen but nearby. The cutting can be set to 25 or less and the cautery to less than 30 to be safe.

Eventually, the silicone coils are appreciated often by carefully visualizing the tissue, sometimes in different lighting or angles during the dissection, where these parts of the lead can be seen nearby, perhaps a layer or two under scarring. Often, parts of the wires within the coils or just proximal to the coils end up being exposed through the silicone. It is important to avoid hitting the exposed wires with cautery energy as this can transmit a significant amount of energy into the nerve. All three coils and their internal wires can generally be removed in this process, sometimes gently pulling the coils through the scar if they are close enough. In this manner, the coils can sometimes come out in pieces but eventually allow for the entire lead to be removed.

Dissection then continues working closely along any parts of the nerve that can be appreciated and freed up further. Enough of the nerve for a new lead can usually be safely freed. Often there remain areas of scar that seem to be unsafe to remove from the margins of the nerve. In a replacement, it is often acceptable to place the new lead over most of these as long as the bulk of the scar has been removed. The new lead is then placed in a typical fashion, secured by the last anchor coil on the nerve. It is then anchored in three locations in the soft tissue after carefully removing the blunt Weitlaner retractors. Once secure, the IPG incision is opened, and the IPG removed after freeing it up with the cautery primarily. The old remaining lead, having been cut in the neck area, is pulled through to the IPG area, and the entire old IPG is either disconnected and reused with the new lead if still viable or is removed with the old wire and discarded. In this case, a new IPG was placed because the last IPG change was already several years prior and there was some concern that the cautery could have damaged it as well. The tunneler was tunneled from the IPG pocket to the neck area, and the lead was pulled through the clear tubing to the pocket where it was connected to the new generator and placed back into the pocket. Testing the system then proceeded in the usual fashion, and all impedances were within a reasonable range. Heart rate detection also was accurate, and the incisions were then closed and dressings were placed. The patient was awakened, extubated, and taken then to the recovery area for further care.

It should be appreciated that many surgeons find this dissection too risky and instead simply cut the lead off as deep as they see is safe to extend the dissection. However, this decision only can be used to manage the elimination of the lead without replacement. If the lead is infected, it does n't remove all the hardware. If the patient needs an magnetic resonance imaging (MRI), they would need to have less than 3 cm of the lead left in place to be safe, and this is difficult to measure when it is still ensconced in scar. As a means of removal and replacement of a lead, leaving a portion of the prior lead would then require that enough vagus nerve be dissected free above or below the remaining coils of the prior lead to place a new lead. This is often possible, although it still requires that the cut prior lead be small enough to still allow for any chance of having an MRI performed if needed. True, complete removal of a stimulator lead is often possible, however, with diligence and care and can proceed more or less in a similar timeframe to the original placement of the lead, with perhaps a few more minutes allotted for the dissection. The difference, primarily, is that the original placement requires a dissection to find the vagus nerve and free it up for a 3-cm length, more or less. The removal, in contradistinction, proceeds primarily by following the lead down through the scar, discovering the nerve, as such, within the scar and where the coils of the lead reside. Once there, it often becomes more clear how to dissect the nerve free once the silicone coils are removed.

> <u>Variation</u>: *The lead can't be removed completely* – There is a point in any of these dissections when the surgeon must decide that it is more risky to continue to try to get the entire lead off the nerve and removed than to continue. There is no rule for deciding this, and each surgeon innately knows for themselves where these boundaries lie. If it is the case that one feels the lead cannot be removed, it is best if less than 3 cm of the lead can be left, as this may allow an MRI to be performed still. Attention should be shifted to thinking about dissection of the nerve to place a new lead if that is the plan—often this is done by moving the focus of the dissection region further superior. In some cases, the rest of the lead eventually can be removed once the nerve is dissected further from this other area, so that possibility should still be kept in mind. If the lead cannot be removed and no new lead is to be placed, it is reasonable to proceed to closure.

> <u>Variation</u>: *The nerve cannot be dissected fully enough to place a new lead* – In cases wherein a new lead is needed, it is usually possible if the entire old lead can be removed as this necessitates understanding the anatomy of the nerve and scar and surrounding tissues and often is adequate for enabling necessary nerve dissection. However, in some cases, of course, the nerve will not be able to be dissected free enough for new lead placement. At some point, this determination must be mad, and either the surgery is abandoned at that juncture or a decision to close and place a lead on the opposite side (i.e., right side of the neck) should proceed in a timely way. The important aspect of these surgeries is that the patient (and/or caregivers/power of attorney/health-care proxies) has been apprised of this possibility and options ahead of time.

Chapter 22

Evaluation and programming in vagus nerve stimulation (VNS)

Patricia Osborne Shafer[1,2]

[1]*Health Information and Resources, Programs and Services, Epilepsy Foundation, Landover, MD, United States;* [2]*Neurology, Beth Israel Deaconess Medical Center, Boston, MA, United States*

Scenario

A 33-year-old female with drug-resistant focal epilepsy presents for consideration of a vagus nerve stimulator to aid in seizure control. Seizures with change in awareness began at age 20, although upon questioning, she likely had focal aware seizures starting in early teen years. She has tried multiple antiseizure drugs, with some improvement. She continues to have focal impaired aware seizures 2 to 4 times a month and focal impaired awareness every 2–3 months. She had to change careers because of cognitive impairment from seizures and medication side effects. Her inability to drive has limited employment options. She is married and would like to have children. She is interested in exploring all options to improve seizure control prior to pregnancy. Several discussions ensue regarding a plan for further trying to eliminate her seizures.

Resective epilepsy surgery typically offers the best outcome for seizure control, if seizure onset can be localized to one focus in an area that could be safely removed. Video electroencephalography (EEG) monitoring that was done about 5 years ago at a different epilepsy center could not be obtained in this case unfortunately, except for several reports of the interpretations. The patient is not clear if both seizure types were recorded or if medications were withdrawn during monitoring. We recommended repeating the video-EEG in-house monitoring for more definitive information on her seizure onsets.

> **Variation**: *The patient does not want to pursue further testing* – While repeat testing to consider all treatment options is the preferred choice, patients may be reluctant to consider invasive surgery because of work insecurity and desire to have a child, or other reasons. Even if results suggested a cortical resection, patients may prefer to pursue a less invasive approach first. Vagus nerve stimulation (VNS) could be offered in this setting at this juncture of the workup if that were the case.

The Neuromodulation Casebook. https://doi.org/10.1016/B978-0-12-817002-1.00022-5

Scenario (continued)

She was referred for reevaluation of surgical therapy before scheduling VNS surgery. She underwent video-EEG monitoring which found bitemporal seizures on surface monitoring. Tachycardia was noted prior to the focal impaired awareness seizures. Research has suggested heart rate change may occur prior to 80% of seizures. Magnetic resonance imaging scan was normal. The epilepsy team recommended invasive EEG monitoring to see if most or all of her seizures arose from a single area. If resective surgery was not an option, the possibility of implanted devices could be considered. The patient remained reluctant to undergo surgery for implanted electrodes or any open craniotomy given her concerns about time off from work and desire to get pregnant. She was seen again by neurosurgery and epilepsy nursing to discuss vagus nerve stimulation. Given the tachycardia noted prior to some of her seizures, the responsive feature of VNS with heart rate change could provide additional benefit to the most disabling seizures. Surgery was scheduled and performed (see Chapter 22).

> **Variation**: *The patient is concerned about potential side effects of VNS therapy* – Side effects from placement of VNS are infrequent and include potential for infection and vocal cord paralysis. Side effects during programming relate to stimulation and usually lessen or resolve over time. The most common adverse effects during programming include hoarseness, tightening or funny sensation in throat, cough, difficulty swallowing, or sensation of breathlessness. Often severity or presence of side effects can be managed by changing stimulation parameters. For example, lowering pulse width may lessen discomfort/pain in throat area. Lowering output current or signal frequency may reduce hoarseness. Adjusting the duty cycle to shorter on/off times may also help manage adverse effects. Although patients can be apprised of all these potential side effects and although they typically are all treatable and do not eliminate the ability to use VNS as a therapy, patients may still be hesitant to pursue surgery at all for their condition.

Scenario (continued)

She is brought to surgery on the elective schedule (see Chapter 22). Diagnostic testing during placement of the VNS generator with SenTiva Model 1000 and lead was uneventful. Lead impedance was acceptable at 1095 Ω with good communication. Heat beat detection was stable when set at 3 with 40% sensitivity. As this was her first VNS, output currents were set at 0, with heartbeat detection turned off after satisfactory intraoperative testing.

The patient was seen in epilepsy clinic 2 weeks after VNS placement to begin programming the device as is standard. Programming the device earlier is counterproductive, typically, as the impedances are changing more rapidly in healing and fibrosis and amplitudes would likely need to be adjusted every few days. Waiting 2 weeks to start usually eliminates these problems. She had two seizures since surgery, one focal aware and one focal impaired awareness seizure. Incisions were well healed without signs of infection. No hoarseness or other adverse events were reported.

The standard programming algorithm was started. Programming of VNS can be done using a standard approach, increasing output current by 0.25, with pulse width at 250 or 500. The Model 1000 allows for increases by 0.125 increments for output currents initially. This enables the programmer to set autostimulation output currents differently from normal mode parameters. However, less flexibility is available for pulse width parameters. Personally, I have found side effects more bothersome initially with a pulse width of 500 Hz. While many people may experience side effects for hours after the stimulation is begun, it usually resolves within hours or over the course of 1–2 days. She seemed to tolerate this initial programming reasonably well. She is scheduled to be seen in the epilepsy clinic every 2–3 weeks for programming over the next 2 months. This allows the health-care providers to assess tolerance to side effects and make slow titrations to minimize bothersome symptoms. Tailoring often requires adjusting one or more parameters (output current, pulse width, and/or on/off times) depending on presenting symptoms and tolerance.

> <u>Variation</u>: *The patient indicates it will be difficult to get transportation to clinic for follow-up visits* – The SenTiva Model has the capability to begin programming in the clinic setting and preprogram subsequent changes to occur at home at preset intervals. This can be helpful for people who have difficulty with transportation to appointments. Unfortunately, the changes would occur without an assessment by the health-care provider. Thus, if a person has difficulty tolerating preset programming, it cannot be adjusted unless the patient physically comes into the epilepsy clinic.

Scenario (continued)

She returns for a single adjustment visit where the amplitude only is increased by 0.25 mA. However, the following month, she presents to our epilepsy clinic reporting breathlessness and throat tightness when exercising and with magnet use. Mild hoarseness also occurred when the magnet was used for a recent seizure. Of note, magnet and autostimulation settings were set higher than normal mode parameters. Her seizure frequency seems to have decreased, nevertheless, with only one seizure in the prior 2 weeks. There are several variations and information that may contribute in handling this situation.

Variation: *Continue standard programming* – As side effects resolve in most people over time and she had another seizure since last visit, standard programming settings could be continued but increased by lower incremental amounts, such as 0.125 mA instead of 0.25 mA. Care should be taken with autostimulation and magnet modes until side effects can be monitored further. As she has noted side effects with exercise, encouragement is given to her to avoid this or change the type of exercise until side effects resolve.

Variation: *Turn off or lower stimulation temporarily* –When side effects are bothersome, some people may prefer lowering one or more parameters and wait another few weeks before retrying upward titration. Given her recent seizure, this could delay any positive impact on seizure control. Additionally, stopping or lowering stimulation amplitude does not help sort out which parameters, if any, may be contributing to the side effects.

Variation: *Tailor programming to tolerance, comfort, and efficacy* – Upon questioning, symptoms seem to be happening when working out at the gym and once when the magnet was used. Her program history shows some stimulations in response to heartbeat, but she was not tracking dates and times of her exercise program to identify patterns yet. Adjusting the sensitivity of responsive stimulation from 40% to 60% and maintaining settings at current levels may prevent exacerbation of side effects while exercising until she gets familiar with her settings. Additionally, a patient can track activity to compare side effects and stimulation changes. Heartbeat sensitivity and other parameters can then be titrated as she acclimates to the stimulation.

Scenario (continued)

She was able to tolerate programming without further difficulties when a tailoring approach was used to minimize adverse events. Heartbeat detection was used with sensitivity lowered to 40% again. She was then able to resume exercising without difficulty. Seizure frequency decreased by 25% in the first 3 months and by 50% after 1 year. A review of her programming history recorded a number of activations in response to heart rate changes.

The duty cycle was increased gradually to ∼30%, and her seizures changed to primarily focal aware seizures every other month with no focal impaired awareness seizures. She was working consistently and eventually gave birth to her first child. By 3 years after implantation, however, her seizure frequency started increasing again to monthly focal aware seizures and two focal impaired awareness over a 2-month period. She called in and reported these seizures, and we promptly asked her to come into clinic. We found during diagnostic testing that there was a lead impedance high at 5000 Ω. Several variations could be considered at this juncture.

Variation: *Adjust anti-epileptic drugs (AEDs) and watch* – Blood tests to assess changes in AEDs and rule out pregnancy can be performed. In this case, her pregnancy test was negative and AED levels had not changed in the previous 6 months. As she is a new mother with many potential stressors, she was cautioned to avoid sleep deprivation and approaches to manage stress and other triggers. She was scheduled then to follow up again in 1 to 2 months and consider medication changes if the increased seizure frequency persisted.

Variation: *VNS programmed in response to improve seizure control* – VNS stimulation parameters were reassessed. Her pulse width had been kept at 125 μs over the past year. Increasing pulse width to 250 or 500 μs may improve seizure control. However, this could trigger side effects that she experienced early in course of VNS. She should be seen in a few months to reevaluate. Other parameter changes could be considered. Duty cycle could be increased, but at certain levels, the autostimulation mode would be inactivated. Another option would be to leave the duty cycle stable but adjust heartbeat (HB) detection sensitivity. The impact of these changes on battery life should be considered.

Variation: *No change in parameters but obtain estimate on battery life* – Some stimulation parameters may have greater impact on battery life than others. While reported battery life of the devised used is about 10 years when standard settings at 10% duty cycle are used, a higher duty cycle or frequent autostimulations can shorten the battery life. The lead impedance has increased to 5000 Ω, which itself is suggestive of a shortening battery life. As seizures are increasing in frequency as well, further analysis of battery life is indicated. If seizures persist or adjusting parameters does not help, prophylactic replacement of the battery should be considered. Alternatively, the epileptologist could consider adjusting AEDs during this time.

Scenario (continued)

She feels that the device has been very helpful to this point. She would like to know if the device should be changed and is willing to consider medication changes while waiting to see if seizures worsen either in frequency or severity. However, she does not want to pursue medicines that would increase her side effect burden. Further evaluation does not show evidence of a lead fracture.

Over the subsequent few months, seizure frequency does gradually increase. Battery life estimates from the manufacturer suggests less than 1 year of remaining battery life. The patient and the epileptologist both request a replacement of the VNS generator. This is performed within the next 6 weeks, and although the settings are otherwise maintained, the amplitude is started at 0.125 mA below the prior setting in the immediate postoperative period because of the fact that the waning battery life may have led to over-amplification as a compensation. She was followed up for a few weeks with no increase in seizure frequency. The increasing seizure occurrences seemed to have been stabilized. Her amplitude was increased by 0.125 mA again, and her seizure frequency decreased over the subsequent 6 months to almost zero without any medication adjustments.

Chapter 23

Responsive neural stimulation for epilepsy

Jeffrey E. Arle[1,2]

[1]Neurosurgery, Harvard Medical School, Boston, MA, United States; [2]Neurosurgery, Beth Israel
Deaconess Medical Center, Boston, MA, United States

Scenario

A 27-year-old man who had had seizures for over 15 years was being evaluated for possible invasive monitoring to define his epileptic focus in a more localized fashion. His hope, of course, was that a resectable focus could be found, and he was looking forward to proceeding with that surgery in the near future after several years of deliberation over this option. He had had several discussions with his parents and his epileptologists (two different neurologists over the past 10 years) regarding potential risks and benefits of intracranial epilepsy surgery. One reason he decided to move forward now was that he had recently had an increase in his seizure frequency, despite recent pharmacological attempts to regain some modicum of control over them again. For a few years in his early twenties, he had only one or two seizures per year. However, more recently, seizures had increased in frequency, and in the last 6 months, they had become now almost one per week, making his ability to maintain the employment he had in a start-up company as a programmer very difficult. Although supportive, his superiors indicated they hoped he would be able to return to work within a 6-month timeframe if they gave him time off to try to resolve things better with his epilepsy.

Another aspect of his circumstances, given he had an increase in seizure frequency, was that he now was aware that even if a resectable focus was not able to be found, a new device could be implanted that would simultaneously record brain activity and then, if a seizure was noted to be starting, stimulate in an appropriate manner to minimize or stop the seizure from progressing and propagating. This was known as a responsive neural stimulation (RNS) system. This gave him an extra shot at a reasonably good outcome, although the likelihood of becoming seizure free was still fairly low without a resectable focus per se.

The Neuromodulation Casebook. https://doi.org/10.1016/B978-0-12-817002-1.00023-7

Although he had had several extracranial monitoring sessions through the epilepsy center and two magnetic resonance imaging (MRI) scans over the most recent 5 years (one with a specific epilepsy protocol), no lesion (such as a region of cortical dysplasia) had been found nor had a clearly reliable onset location of his seizures been located. One problem was he had two seizure types: an infrequent generalized tonic-clonic pattern, which he had n't had in over 2 years at this point, and the second (now increasing in frequency) which typically began with an interruption of attention or speech if talking. These would resolve within 10−15 seconds most of the time but would progress with eventual head turning and arm rising on the right and lip smacking, with occasional vocalizations in less common incarnations. Some of these seizures occurred during sleep, although it was unclear how often he had nocturnal seizures. Monitoring had localized onsets to the right side but could not isolate temporal onset possibilities (either mesial or extramesial with rapid mesial spread) from potential frontal onsets, which had been more apparent in some of the recorded seizures at times. An intracranial study would be needed to better determine the actual onset, if possible, and discussion site centered on the plan for electrode placement in that initial operation.

Part of this plan included the realization that electrodes would be needed over both temporal and frontal areas and a larger grid would be necessary for more definitive mapping of cortical regions, as well as a depth electrode in the both the right and left hippocampus. As such, the surgical plan would require a craniotomy instead of just bur holes or small craniectomies to place strip electrodes and a few depth electrodes. If RNS was to be considered in the future, the ability to implant the device in the remaining area of the skull might need to be modified. In addition, a sodium amytal test was performed, which allowed the consideration of temporal and mesial temporal resection if necessary. A 256-channel high-density extracranial monitoring session was performed as well, and the more recent 3T MRI was rereviewed.

Variation: *Owing to previous surgery, there is not enough room to place the responsive neural stimulation device on the same side of the head* – It may be clear that because of difficulties healing from prior surgery or the location of the previous incisions, the new incision to place the RNS device cannot be made reasonably on the same side of the scalp. Incisions may compromise blood supplies, compromising healing, or the curvature of the open areas on the scalp for an incision and the curvature of the device itself do not match well enough. Typically, there is only a region of the parietal area that fits well enough in this regard. Sometimes the region slightly more anterior on the skull can be used as well, spanning from parietal to frontal regions. In these cases, plans must be made to accommodate the device on the opposite side of the head. Placement of the head in the Mayfield head holder and its positioning in the operating room (OR) for surgery needs to be carefully considered in such cases, allowing access for the electrode placements and their tunneling to the device. An alternative option to be considered in certain cases as well is to place the device wholly within a prior craniotomy, the opening for the tray being well within the margins of the prior bone flap size. This option works out well when it can be accomplished.

Scenario (continued)

He passed the sodium amytal testing, which supported the possibility of a right hippocampal resection if necessary and also indicated that his language function was on the left side. The 256-channel study did not localize his seizures any better than prior monitoring sessions. There were three seizures, all of which occurred after some tapering of one of his medications and which showed what appeared to be right temporal onset but very rapid spread to frontal areas. However, one of the three onsets was nearly identical in time to right frontal onset, indicating that the focus could be deep to the cortical outer surface. After further discussion of these findings and the findings all together, it was decided that he would undergo an invasive study with a frontal grid which extended over part of the superior and middle temporal gyri, with subtemporal strips and mesial and subfrontal strips as well as some frontal opercular depths and a hippocampal depth electrode. The craniotomy would likely allow for the placement of the RNS device just immediately posterior to the craniotomy site if necessary.

The patient arrived a little nervous for the first surgery to place the invasive electrodes. He was reassured and had the expectations redescribed to him regarding being in the intensive care unit afterward with a large headwrap and protection in bed to keep him from accidently pulling out electrodes during a seizure. The operation went reasonably smoothly with the right-sided craniotomy made with a typical reverse question mark incision and using a navigation system to help place the hippocampal depth electrode from an occipital approach as well as the several frontal depth electrodes and the planned grid and strips in the frontal and temporal regions.

> **Variation**: *Prior surgery makes it difficult to place strips and/or depths* – It may be that scar tissue or adhesions and critical tethering of important blood vessels nearby or eloquent brain regions at risk with dissection (language areas trying to get to the insular region, for example) limit the ability to place strips or depth electrodes appropriately. In general, if a monitoring strip or grid was able to be placed, then the RNS electrodes can be placed in the same region (taking care to leave markers if required that are clear for indicating the plan). However, when returning in the future to place the RNS, difficult dissections

may be required. A backup plan for electrode placement should be thought through ahead of time to manage such restrictions. The backup plan may well be to abandon the placement of the RNS device, and the patient should be apprised of this possibility prior to surgery.

Scenario (continued)

Although he was sluggish in the first 2 days after this initial surgery to place electrodes, he eventually came around well and was at his cognitive baseline. Computed tomography (CT) and MRI scans as well as anteroposterior and lateral skull films were obtained prior to connecting all the electrodes into the recording system by the technicians. Minimal evidence of any cerebrospinal fluid (CSF) was appreciated during the early phase of the monitoring period during the first week. No seizures were reported during the first 5 days, unfortunately. After some medication tapering, some right midtemporal spiking was appreciated but no seizures. The medications were fully tapered, and some sleep deprivation medications were added when he had three seizures in 1 day. In fact, after the third seizure, Ativan was given to prevent him from going into status epilepticus. Ultimately by day 10, he had had seven seizures, six of which were corroborated on video and one which, unfortunately, was not captured by video as there was a disruption in the link at the time. However, further analysis of these seizures led to the conclusion that onsets could still not be corroborated between an early mesial location with rapid spread to frontal and extramesial temporal areas and a separate focus in one of those areas, primarily frontal under three of the grid contacts. By the end of the monitoring period, several changes of the headwrap, however, were now required because CSF drainage was increased out of one corner of the incision and the dressing had become saturated. The area was reinforced and rewrapped, but there was concern for an increased infection risk in this context.

Because onsets were likely temporal but with such rapid spread to frontal areas or frank onsets from there, overall success in eliminating seizures from a right temporal lobectomy with hippocampectomy was deemed unlikely. It could be tried, but becoming seizure free in the long term (>2 years) was thought to be below 40%. A second option of using RNS was put forward, wherein recordings from the right hippocampus could be performed using a depth electrode placed there and stimulating electrodes placed over the area of the frontal lobe, with a second strip electrode left over the suspect area of the temporal lobe cortex which had early activity as well, but not connected into the device. After a moderately lengthy discussion with the patient and his parents, the patient decided to opt for the RNS implant.

Given the concern about CSF drainage seeping through the headwrap during the 2 weeks of monitoring and that the white blood cell count had started to rise in the final 3 days as well (although he had no fevers or other signs of an infection), the decision was made to remove the electrodes and leave markers for the RNS placement later. This would be an attempt to avoid an increased risk of infection for this elective implanted device. Some data, however, have suggested that the infection risk of placing the RNS at the time of removal of monitoring electrodes creates no increased risk of infection. The patient was taken back to the OR after disconnecting all the electrodes from the monitoring cables and restarting all of his preoperative medications. He had the craniotomy site reopened and all of the grid, strip, and depth electrodes removed while carefully marking the brain surface underlying the appropriate contacts on the electrodes. A photo was taken of the field before electrodes were removed for future reference. Important or likely onset areas under certain contacts were marked using paper ruler numbers cut in small squares right around the numbers and carefully placed with forceps onto the brain surface before removing the electrodes.

> **Variation**: *Important cortical regions are located outside of the cranial opening* – This is not uncommon if the electrodes were placed through a craniotomy but were slid into more distant areas for coverage. Even a large grid often extends under the edges of the cranial bone flap opening. Thus, the possibility of having to account for the correct trajectory with the RNS electrodes when one cannot be sure of exactly where the RNS electrode sits is a real one. Most of the time, the full extent of the electrode can still be seen—the trick being in securing it so that it remains in place there. In less typical circumstances, visualization of the electrode is not possible. It is critical that effort be taken to make sure one understands the true orientation of the head in space and in the holder, although it is under the drapes at the time, and the actual trajectory of the strip or depth electrode being placed cannot be fully appreciated. It is not difficult to interpret one's place on the cortical surface to be slightly different than reality and place a strip electrode 30 or 40 degrees off from the intended location.

Scenario (continued)

After this, the planned trajectory of the RNS strip electrodes was considered. Microscrews were then left in the bone margin edges in such a way so as to indicate what the plan for direction and location of the RNS electrodes would be when coming back to place this device in the future. Another photo was taken, and an appropriately detailed description of this was left in the medical record as well.

> **Variation**: *What about placing the responsive neural stimulation device at the time of electrode removals?* – Ideally, one can place the RNS device at the time of surgery for removal of the monitoring electrodes. This can be considered in most cases. If there are concerns

about infection, then placement should wait a couple of months. If the patient has not yet decided on whether they want to move ahead with RNS, then placement would need to wait as well, although it would be important in such a case that marker screws be placed anyway at the time of the electrode-removal surgery.

The patient recovered reasonably well from the electrode removal and remained in the hospital for only a day before being discharged home. No seizures were reported during the next 3 weeks, during which time he returned to have staples removed and had healed well. He then returned after about 2 months to have the RNS placed. There had been no evidence of infection, and he had not been given any antibiotics during this time. He did report three of his typical nongeneralized seizures, however, which occurred after going for the first 6 weeks postoperatively without any seizures.

After reviewing the risks of the RNS placement (infection, stroke, movement of the leads, hemorrhage, no benefit for seizures, erosion or poor healing over the device), the patient was brought into the operating room, intubated, placed in the Mayfield head holder, and had the navigation system initialized based on their prior MRI so that the hippocampal depth electrode could be placed using the articulating stereotactic arm. This method eliminates the need to place the patient in a stereotactic frame first for the longitudinal depth electrode placement, followed by the need to remove the frame and then perform the craniotomy with new prepping and draping. As such, it also eliminates the need for a scan in the frame preoperatively. The head was shaved, prepped, and draped for an occipital incision on the right, placed and aligned to place the hippocampal depth electrode longitudinally from the occiput. Also, the prior well-healed anterior frontotemporal question mark incision was to be reopened, and a new parietal region inverted "U" type incision was to be made a few centimeters behind the margin of the prior incision to allow placement of the RNS device itself. As such, he was asked whether it would be ok to shave all the hair, and he agreed, allowing it all to grow back in equally. This made planning the incisions easier to see and for dressings to stay in place better afterward.

The surgery began with a small incision in the occipital area after targeting the hippocampal trajectory using the navigation and adjusting the articulating arm with a holder for the cannula. A standard 14-mm bur hole was placed where the entrance point was planned, and the cannula was slowly advanced to the target. The RNS depth electrode was then advanced through the cannula after removal of the stylet, and then the electrode was carefully held in place until the cannula was slowly removed from the brain, leaving the electrode in position. The electrode was secured to the nearby galea with a 2-0 silk suture and a small anchor surrounding the electrode. The electrode was carefully left dangling nearby still sterile, awaiting tunneling to the RNS device at the parietal incision location. A small piece of gel foam was placed over the bur hole and where the electrode exited from the brain. Attention was then turned to the prior question mark incision. This was reopened using a scalpel and standard technique for performing a craniotomy. The prior bone flap was removed, and the previously placed markers indicating locations for the electrodes were found and noted. The prior placement plan was recapitulated from the locations of the marker screws and prior notes in the records to indicate how the strip electrodes should be positioned. The two 4-contact strips were placed carefully in those locations after making sure adhesions and scar tissues were removed, and they were then secured to the dural margin using 4-0 Nurolon. It was necessary to secure one of the strip electrodes with two 4-0 Nurolon sutures to prevent windshield-wiping of the lead after closure of the dura. Once secured, after irrigation with saline, the dura was closed in typical watertight fashion, with the wires from both electrodes coming through the dural incision but with sutures securely tied right around those areas to prevent CSF leaks.

At this juncture, the RNS device incision was made in the parietal region as planned and hemostatic clips were applied to the scalp and galeal margin in typical fashion. The device template was used to carefully mark the skull in the outline of the device, and then standard bur holes were made in several locations inside the edge of this outline. The craniotome was used to connect them and follow the outline as exactly as possible. The slight differences were shored up with the typical matchstick drill bit, repetitively testing the fit of the tray that would hold the implantable pulse generator (IPG). Once bone had been removed in this manner so that the tray fit the opening exactly, the tray was secured by screwing the flanges to the bone edge and then placing the IPG into it. The leads were tunneled to this opening using the provided tunneling tool, and then the strip and depth electrodes were connected and secured in the manner described by the representative from the company who was in the OR to help make sure connections were made properly. The second strip electrode was left nearby with a special covering over the end of the lead to protect it in case it was to be used in the future. Once connected and impedances all tested within normal variances, all three incisions were closed in standard manner, and dressings were placed before taking him out of the Mayfield pins. A secondary headwrap of gauze was then also applied, to be removed in a few days to minimize swelling in the galeal layer.

His postoperative course was reasonable, and he was able to be discharged with only a modest amount of pain the following day, after MRI and CT scans had been performed. These scans allowed the appreciation of where the electrodes actually were placed relative to the planned locations and the seizure foci as mapped in the prior surgery and monitoring. Programming would take place over the ensuing months, initially passing through an extensive period of recording and linking any seizures recorded to waveform analysis to allow programmable algorithms within the device to appreciate when a seizure for him was starting. Following programs would then send stimulation to nearby contacts on the electrodes to disrupt the continuation and spread of the seizure. In general, it could take several months to begin reasonably programming him to provide benefit. Often, such programming is highly dependent on how many seizures are recorded and how consistent they are to each other to enhance recognition and targeting of stimulation in this closed loop system.

Index

Printed in the United States
By Bookmasters